THE INTERNET
OF ANIMALS

MARTIN WIKELSKI
Foreword by KEITH GADDIS

THE
INTERNET
OF
ANIMALS

DISCOVERING
THE COLLECTIVE
INTELLIGENCE OF
LIFE ON EARTH

GREYSTONE BOOKS
Vancouver/Berkeley/London

Greystone Books Ltd.
greystonebooks.com

Cataloguing data available from Library and Archives Canada
ISBN 978-1-77164-959-9 (cloth)
ISBN 978-1-77164-960-5 (epub)

Editing by Jane Billinghurst
Copy editing by Dawn Loewen
Proofreading by Jennifer Stewart
Indexing by Stephen Ullstrom
Jacket and text design by Belle Wuthrich and Jessica Sullivan
Jacket photograph by Martin Harvey/Alamy Stock Photo and EcoPic/iStock Photos
Interior illustrations by Javier Lazaro
Interior photographs by Martin Wikelski, except 1 (bottom) Bernhard Gall;
2 (top) Bill Cochran; 3, 4, 7 (top and bottom) and 8 (top) Christian Ziegler;
5 (top) A. Catherine Markham; 5 (bottom), Sergio Izquierdo; 8 (bottom) Uschi Müller

Printed and bound in Canada on FSC® certified paper at Friesens. The FSC® label
means that materials used for the product have been responsibly sourced.

Greystone Books thanks the Canada Council for the Arts, the British Columbia Arts
Council, the Province of British Columbia through the Book Publishing Tax Credit,
and the Government of Canada for supporting our publishing activities.

Greystone Books gratefully acknowledges the xʷməθkʷəy̓əm (Musqueam),
Sḵwx̱wú7mesh (Squamish), and səlilwətaɬ (Tsleil-Waututh) peoples on
whose land our Vancouver head office is located.

For Fiona, Laura, Larissa, and Uschi

Contents

Foreword

IT TAKES A SPECIAL SORT OF PERSON to envision, create, and execute a large-scale, groundbreaking scientific project. Martin Wikelski is such a person. ICARUS, the project he has skillfully shepherded into existence despite obstacles that would have derailed many researchers, has changed the way we view our planet.

Good science that helps us solve pressing problems requires not only solid research but also innovative thinking and a willingness to embrace the possibility of what many would consider impossible. How do scientists dream up and plan for ideas no one has tried before? How much do they rely on groundwork by those who have gone before them and how much do they strike out on new paths? How do they deal with setbacks?

The story of ICARUS lays out Martin's voyage of discovery. His love for the planet and the animals that live here shines through on every page. He is interested in everything from the biggest picture imaginable—the survival of this planet using the collective intelligence of animals—to the tiniest details, such as how to track a dragonfly. He marvels at the feats of migration undertaken by small birds and the way it sometimes seems that animals are the ones domesticating us and teaching us how to play rather than the other way around.

A hallmark of a successful project leader is their ability to build teams and get people on board with the mission. The stories in this book of Martin working with everyone from a single researcher in the back of a small car racing around grid roads in the American Midwest with a radio antenna sticking up out of the roof, to committees meeting at the Russian space agency, Roscosmos—where the smell of communism was still "hanging in the curtains"—show he is as comfortable supporting dedicated researchers as he is navigating the convoluted corridors of power.

Martin's enthusiasm is infectious and his dreams for a better way to approach biological diversity using our capabilities in space are irresistible. Along the way, he never loses sight of the amazing capabilities of the animals he interacts with, from a young sea lion who seeks refuge in his tent in the Galápagos Islands to avoid being caught by the big beachmaster bull, to Hansi, who missed migrating with other storks, adopting a local farm family who fed him finely ground meat and offered him warm foot baths in the chill of winter in southern Germany.

I first became aware of Martin's work through his collaboration with multiple NASA scientists to expand the ability of satellite data to track and predict animal movement patterns. Through his efforts, we are entering an age where migratory patterns of species around the world will be translated into regular mapped products like wind and ocean currents. Martin's work has already led to revolutionary advances in understanding the relationships between natural and human-induced processes and animal movement, behavior, physiology, and health. More importantly, the capability he has enabled is used to direct conservation actions protecting threatened species and habitats across the globe. His vision has turned science fiction into reality for the benefit of life on Earth.

KEITH GADDIS, NASA, Biodiversity and Eco Forecasting Program

Prologue: A Sea Lion Named Baby Caruso

THE SEA LIONS on the beach are loud today. Something must be going on. I'm sitting under a wide shade tent supported by bamboo poles—our home away from home here on Isla Genovesa—getting ready to cook for our crew of four. We've been living on this uninhabited island in the middle of the Pacific for about five months observing marine iguanas.

Our goal is to find out why the marine iguanas on this island in the Galápagos have a specific body size. Why are they as small as they are and not, say, fifteen times heavier like their relatives on the neighboring island of Fernandina? To carry out our studies, we have set up camp on a beach tucked away on the west side of the island. It's surely one of the most beautiful places in the world, and perhaps one of the most remote. There's nobody here except for the four of us, and we have a special permit to camp on the island. It's a great privilege to spend time in an area where none of the animals are afraid of people, in a place where, it seems, no one has scared or persecuted them for millennia.

Our daily routine consists of getting up at sunrise, emerging from our individual tents, walking to the beach for a morning wash, returning to our communal shade tent to make coffee, grabbing our binoculars and notebooks, and heading out to the beach to observe the iguanas. There are lots of other animals in and around camp: red-footed boobies, blue-footed boobies, frigatebirds, mockingbirds, hermit crabs, sea lions. In fact, the hermit crabs have just pre-cleaned the dishes that I am supposed to wash on the beach today. I grab the plates and pots and cups, put them in a bucket, and walk barefoot over the sand toward the ocean.

As I start rinsing the dishes in a convenient tide pool, I hear calls from a sea lion I don't recognize. Like everybody in our team, I know what each one of the roughly forty sea lions on the beach sounds like. The big bull, our beachmaster, makes deep, growling noises befitting an old bear of the ocean. The sounds I'm hearing today are different, higher pitched and clearer than the others. I track them down to a newborn pup. The calls are simply beautiful and make me happy as I rinse the dishes in the calm pool in preparation for a thorough cleaning in the waves dashing onto the beach.

As I walk over to the ocean, the big beachmaster, who recently lost one eye in a fight with another bull, lumbers toward me with astonishing speed. A quarter-ton blob of muscle and fat in fast motion—heading in my direction.

Perhaps he hasn't recognized me? All I can do is kick sand at him and shout at him briefly. Thankfully, he stops immediately. I'm not a threat to him. It could be that he just mistook me for a young competitor. Or wanted to show me who owns the beach. Just in case. He looks at me, barks a few times, and everything is good. He knows me, we know each other—and we both respect each other's duties. I walk back to camp with the clean dishes and tell my friends about the sea lion who was born with a beautiful voice. We call him Baby Caruso.

Three years later, in spring 1993, we're wrapping up our observations on the island. It's the typical scene: red-footed and blue-footed boobies hanging out in camp, a short-eared owl in the bush behind us hunting a mockingbird, and hermit crabs still pre-cleaning the dishes. After a good day of observations, I retreat to my personal tent as the sun sets to enter data in my laptop computer.

This year, I'm living a luxurious life in camp: I brought along a small table so I can work in my own tent, where I won't be disturbed by the wind and the waves. The entrance faces away from the beach, so my laptop is somewhat protected from the salty air. I'm fully immersed in my work when I hear the beachmaster bellowing. It's still the same old male—the one who lost an eye three years ago in a fight and who is now also getting hard of hearing, it seems. We know this because more often than in previous years, when we wash the dishes on the beach, he appears to not recognize us and comes charging over, stopping only when we talk to him very loudly. He then realizes it's us and not another male sea lion, and he's content. He knows that he's not allowed to enter our camp (because he is a bit too stinky and would trample over all our boxes and dishes and perhaps even damage the ham radio we use to communicate with the outside world).

Beyond the bellowing of the bull, I hear calls I haven't heard for two and a half years: the unique voice of Baby Caruso. It's much deeper now, but unmistakable. I'm thrilled to hear him—and totally amazed. I thought after he was weaned from his mother and left our beach, he had probably died because he never came back to visit. But here he is back on the beach, serenading the females and trying to challenge the old beachmaster.

The old beachmaster is not impressed and makes a mad dash for Baby Caruso. I hear sand being thrown up in the air, a brief fight, deep growling from the beachmaster, and a shriek of fear from Caruso, followed by the lumbering gallop of two sea lions. Sea lions only gallop when they are terrified and urgently need to escape (or when they are in hot pursuit of a terrified opponent). Apparently, the young male has gone too far approaching the beachmaster's females and needs to make a speedy exit.

I hear the sound of galloping sea lions coming closer. Now I'm worried there's going to be a fight in front of my tent, which could be dangerous, like when two enraged dogs get into a fight and totally ignore everything around them. Before I can get up from my table to see what's happening, I spot Caruso lumbering up to the entrance of my tent. He stops abruptly a few feet from me, head up, looking directly into my eyes. Then he lowers his head and sneaks into my tent and under my table, laying his head on my feet. He's not moving at all now and I can hear his fast, shallow breathing. He's totally exhausted. And I'm totally flabbergasted. The beachmaster is close behind him. I can hear him drag himself up to the tent entrance. But he knows he's not allowed here. I shout at him. He recognizes my voice and probably expects me to kick sand at him. He retreats to the beach, still roaring with rage. Caruso remains motionless under my table.

I can't believe what I've just witnessed. Unbeknownst to me, three years ago a sea lion pup must have observed—and completely understood—my social relationship with the beachmaster. Caruso apparently realized and remembered for three years that the beachmaster is not allowed in camp. Somehow Caruso knew that the big bull and I have an agreement: the bull is the boss of the beach and I am the boss of the camp. Caruso had been gone for two and a half years before returning to his birth colony. When he followed his instincts to approach females, he immediately got into deep trouble with the old beachmaster. When Caruso was out of options and close to being seriously beaten up by the big bull, he decided to run to a place he knew that, he remembered, was off limits to the beachmaster. My tent. Caruso had made a mental connection between his sea lion world and my human world. He understood the intersection between the two and knew how to exploit both.

I have made it my life's work to understand the human-animal connection from the human side of the equation.

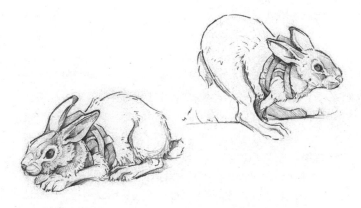

1 From the Prairie to Space and Back

ILLINOIS, TALLGRASS PRAIRIE, 1998. I have just accepted a professorship in the Department of Ecology, Ethology, and Evolution at the University of Illinois at Urbana-Champaign. A group of us are standing in the middle of a huge flat field. It is the end of summer, and the lush green of the amazingly tall grass around us is slowly fading to brown. The impressive growth of the grass is powered by 8 feet (2.5 m) of the most fertile soil on planet Earth just below our feet. In the old days, a person on a horse could disappear while riding through a tallgrass prairie at the end of summer.

This area of the world has also always been a fertile place for novel ideas. The people here are both rock solid and incredibly creative: tradition begets innovation. You first have to master what's out there to invent something fundamentally new. The folks I am standing here with in this field are special too: Bill Cochran and George Swenson are old men now, still squabbling with each other like father and son, or maybe more like older and younger brothers. It is unbelievable what they have witnessed in life, but they are not talking about the past. They are concerned only about the future and where humankind will go.

Forty-one years before, to the day, they stood here in this very field in the middle of the tallgrass prairie. It was during the Cold War. The Soviets had just launched Sputnik. The Western world was shocked. It was the first time that a radio beacon had been put into orbit to fly around the world, or more accurately, to fall slowly down onto our planet. This is actually what satellites do in low Earth orbit when they are not actively powered: there is still enough gravity from Earth and some atmospheric particles left to slow the satellites down, which makes them spiral gradually back to us. And while the world was shocked and paralyzed, George, the older of the two brothers-in-spirit who were nonetheless so unalike, had an idea: Let's build a receiver for Sputnik and listen to the waves coming down. After all, it's just a radio beacon, and all we need to receive its waves is a radio.

George had, until that moment, listened only for other types of waves coming in from outer space: the waves constantly streaming in from the end of the universe from the beginning of time mixed in with all the other radio waves from all the other galaxies out there. Listening to space is just like listening to a concert. The music in an opera by Verdi is not much different from the music from the stars in the universe. They emit waves on slightly different frequencies, but they both give us beautiful symphonies. We just have to tune in. This is what George did all his life. He saw the universe, everything that is around us, as a symphony of waves. This is why he studied music, gunshot explosions, radio astronomy—and Sputnik, when this unique opportunity was laid out in plain sight in front of him.

Sputnik was George's opportunity to make a name for himself in the world of science. All he needed was a radio to receive the waves. So, he turned to his younger brother-in-spirit, Bill. Bill was a mental hippie, somebody who would not do anything others suggested unless he thought it was something worthwhile. He was an off-the-charts genius—according to George, "the most fertile mind I have ever seen," something George told me again much later, when he was over ninety years old, a few weeks before he died in 2017. Bill didn't care whether he became famous or not, but he thought it would be fun to listen to the first satellite.

Bill went down into his basement and did what he had learned to do from his dad: he built a radio to receive radio waves. It was a challenge because the frequency of these waves was tricky to receive, but Bill needed only about thirty-six hours straight to build a radio receiver for Sputnik. Once he had it ready, he called George and they drove out to this very field where they waited for Sputnik to appear above the horizon. And as true as orbit physics can be, Sputnik appeared, and Bill and George, standing at this very spot here in 1957, heard the first human-made noise coming in from outer space.

But that is not the most remarkable thing. They immediately quarreled with each other, even then, and didn't think about what they had just done—listen to Sputnik—but about the future. What could they do with this knowledge? And almost simultaneously they had the same idea. Here's roughly how their thought processes went. Let's first use these signals to determine the precise orbit of Sputnik. And then we'll listen to the signals more carefully and see how they become distorted. Without anything between Sputnik and us, the signal would be perfect, like the A of a tuning fork. But there is the Earth's atmosphere, which changes all the time, and then there's space weather up above and weather on Earth below. But if we were to keep the Earth weather constant, if we were to measure only on good days with similar weather, the signal distortion in Sputnik's beep-beep-beep signal would give us information about compositional changes in the upper atmosphere. How amazing is that! Just these distorted waves—a slightly odd, trembling A note—will give us information about dynamic patterns in the universe. All we have to do is listen carefully. Years later, Bill, George, and I would apply the same principles to listening to animals and the symphony of waves they produce.

Back on the prairie on October 6, 1957, two days after Sputnik's launch, Bill and George took a short break and went home to think about the next steps that would be necessary to measure the upper atmosphere. But when George wanted to check in with Bill a few hours later, Bill was nowhere to be found. George and his friends searched, but nobody could find him. Finally, they opened all the cupboards—and in one of them, they found Bill, sleeping like a baby. He just needed a dark, quiet place to rest after building the first civil Sputnik receiver for thirty-six hours straight.

Within two weeks, Bill and George had all the data they needed first to publish the precise orbit of Sputnik, and then to discuss the composition of the upper atmosphere. But for me, standing with these two amazing but totally unassuming pioneers in the middle of a field in one of the flattest areas of the world, there was something much more important about their work, because once again, we were thinking about the future, not the past. While forty-one years before, George went on to a fantastic career in radio astronomy and Bill founded the field of biotelemetry, their two separate career paths were now converging.

Radio astronomy looks outward from planet Earth to the ends of the universe, often searching for the beginning of time. Biotelemetry, however, looks inward to what is happening right now on our own planet. In biotelemetry, animals are fitted with a device that communicates with a base via a radio transmitter. Biologging is almost the same, but the device attached to the animal generally needs to be removed so the data it has recorded can be manually downloaded and read. Both biologging and biotelemetry are like sending the animals around with a daily diary, a term invented by my friend Rory Wilson, one of the true pioneers of allowing animals to record their own lives. The goal is to visualize the invisible. What are animals doing when nobody watches? And what can we learn from the interactions of all these animals, the most intelligent sensors that will ever exist?

Generating completely novel knowledge is what many field biologists are after these days. Our approach to life (and biology) is undergoing a sea change. We are less interested in the obvious things we can observe with our own eyes or our binoculars or our microscopes. Biologists now want to understand the vast wealth of invisible knowledge animals have that emerges only when we analyze their interactions with each other and with their environment. You could describe this as the search for the sixth sense of animals. It is a grand endeavor akin to the search for gravitational waves, or the last "God particle" (the Higgs boson), or the beginning of time. We will never see any of these realities with our own eyes—we are just not made for seeing them—but we have reached a stage of human knowledge where we can begin to grasp novel realities in both biology (Bill's chosen field) and radio astronomy (where George excelled). And while there is great excitement about where these new findings are leading us in biology, the search for the invisible is a well-worn path in radio astronomy, where many of the first-generation groundbreaking discoveries, such as dark matter and gravitational waves, have already been made.

Standing in that field in Illinois, George told me he felt the glory days of radio astronomy were already over. Initially it had been like many other research fields, where individual universities and research outfits pursued their own projects with little collaboration between institutions. For instance, they all had their own telescopes. They did good research on their own, but a collective of interacting telescopes—which is what we have today, thanks to George—is so much more powerful. We know this from the compound eyes of insects. If you take just one ommatidium, which you could think of as a single mini telescope in the insect compound eye, you can see a little bit. But if you group thousands of them, if you make a composite instrument, you can see everything in incredible detail. Combining telescopes was a quantum leap forward in seeing the universe.

And that is exactly what George did. He led the design team for the VLA, the Very Large Array, on the Plains of San

Agustin, 50 miles (80 km) west of Socorro, New Mexico. Dedicated in 1980, it was the first big telescope array in the world. The VLA is composed of many individual telescopes that all work together to create a common view of the universe. This was a huge step forward for humankind. Despite all their problems with each other, countries around the world were now working together to observe the universe. And George—once again looking to the future—had a few more ideas about what he could do with the VLA.

Extraterrestrials were on people's minds back then, and George was part of a committee on how best to find signs of them. While undertaking amazing observations of the universe, you could also, he decided after thorough rumination, search these data for any signature of intelligent life somewhere out there. And so he came up with the idea of searching for extraterrestrial intelligence as an add-on to the work being done with the VLA. George always impressed upon me that finding signs of intelligent life in space would change our perception of who we are, what we are, and where we are. And for George, it didn't matter if this knowledge came in two years, in twenty years, or in two hundred years. The small amount of extra effort it took to comb through the beautiful symphony of waves coming down from the universe for signs of extraterrestrial intelligence was totally worth it, given the potential for life-changing discoveries.

But Bill, again like a strong-willed younger brother, had his own ideas. He thought that applying the technology he and George had developed for eavesdropping on Sputnik to understanding life on planet Earth would be even more promising and exciting than searching for life in the stars. And George, a few years before he died, confirmed that Bill had been right.

George told me that if he had his life to live over, he would join Bill and the biologists and ecologists to study life on Earth. He told me he thought the emerging field of global ecology was how radio astronomy had been forty years earlier: exciting, transforming, and mind-boggling because of the potential for new discoveries as we looked inward into the most beautiful universe of life right here on our own planet.

So, what did Bill do after listening to Sputnik? Well, it was quite simple. He built a little Sputnik-like radio beacon and harnessed it first to a duck and later to a rabbit. He used a beacon similar to the one he had built to eavesdrop on Sputnik, but whereas Sputnik had sent out radio signals intermittently, Bill built a radio beacon that sent out signals continuously so he could follow his animals without interruption. Once again, he was listening for distortions in the radio signal. Only this time, a change in wave form did not signal a change in the composition of the atmosphere, but rather a change in the respiratory rate of a duck flying out of his hands.

Bill did pretty much the same for the rabbit, except the rabbit was tracked for much longer than the duck. Biologists were baffled by what they learned about an animal they seemingly knew, but really didn't—because this little rabbit did everything differently and unexpectedly when it was out of the biologists' sight. But now, carrying Bill's little Sputnik radio beacon around its neck, the rabbit still talked to the biologists even when they couldn't see it. The waves from the beacon on the rabbit gave the rabbit a voice so it could tell the biologists about its whereabouts, its habits, its fears, and its fun.

These were the two avenues that these men—who shared so much and yet were so very different—took in life. George went to the universe and, ultimately, back to biology, because it is simply more beautiful. And Bill used inspiration from outer space to learn about creatures on planet Earth and, ultimately, to understand the connectivity of life. This connectivity is also what the book you are reading is all about. It is about our connection to outer space and the universe and our human habit of looking away from our most beautiful spot in the universe—and often even forgetting what we have because it's right there in front of us.

2 The Bird Information Highway

BILL, THE YOUNGER BROTHER-IN-SPIRIT, did not rise to the same level of fame as George. But Bill, who died in 2022, may be the one who is saving us now—because he was the one who allowed humankind to listen and speak to animals on planet Earth.

Let me give you an example of how Bill's telemetry connection to animals worked. While standing out in the fields in Illinois listening to Sputnik, Bill was careful enough to listen to other noises in the night sky, as well, and what he heard were the migration calls of the billions of songbirds traveling from the North American continent down to Central and South America. These small birds often live for only a year or two, and in those days they were thought to be guided entirely by their genes. Accepted wisdom was that they instinctively knew where to fly and how many wingbeats it took them to arrive at their winter quarters.

In those days, the actions of small animals were supposed to be predetermined by their DNA. It is a testament to Bill's meticulous attention to detail, to his belligerent, rebellious thinking, and to his natural inquisitiveness that he questioned this on the principle that accepted wisdom is usually at least partially wrong. While listening to Sputnik, he had heard not only individual Swainson's thrushes calling, but also veeries and hermit thrushes. And sometimes he heard a Swainson's call and, astonishingly, a veery answered. Why was that? Was it just a coincidence? Or was there more to it? Could those little songbirds be communicating with each other in the night sky? Instead of individuals following predetermined genetic instructions, could it be more like an Italian highway in the old days, when cars constantly honked to each other to indicate where they were, what they were doing, and the direction in which they were traveling?

While at the time this was a question that could not be answered, when I arrived at the University of Illinois, Bill and I set out to address it together. By the time I joined him, Bill had already been working on the mysteries of songbird migration for quite some time. First, Bill had built an array of audio receivers on the ground. The principle was much the same as the VLA that George had constructed to pinpoint radio waves so he could investigate black holes and how they interact with stars. When songbirds flew close to the microphone, Bill could pinpoint the calls of individual songbirds in the night sky. What he discovered was staggering indeed. The little birds were apparently calling more often when other birds were in the same layer of air—that is to say, flying at the same altitude—and they seemed to be responding to each other. Obviously, this phenomenal finding needed to be tested.

What Bill did next was to broadcast call notes of these bird species from the ground at the exact moment he heard the birds fly by. He found he could entice Swainson's thrushes to respond to Swainson's calls, but even more exciting, he could entice veeries and hermit thrushes to respond, as well. Not always, but much more often than could be attributed to chance. Not only was this an incredibly interesting finding on its own, but even more importantly, it completely changed our understanding and our hypotheses about how these 1¼ ounce (35 g) birds find their wintering grounds.

The constant chirping back and forth in the night sky indicated that even though the birds had some innate tendency to migrate coded into their genes, they still communicated constantly on their journey. An even more radical interpretation of Bill's data was that the only innate tendency the birds needed to have in their genes would be the drive to fly toward warmer areas in fall. The remainder of their intercontinental migration might actually be as simple as following other birds that had made the journey before. All the birds would need to do to find their way south to Central and South America would be to follow others flying along the nocturnal highway.

You will probably not be surprised to hear that Bill was not content with this explanation, either. He wanted to know what exactly these individual birds were doing in mid-air, how much it helped them to call, and what those calls meant to other birds flying with them. What exact information were they communicating? For this, he would need to listen to individual birds as they made their nightly journey, which was an impossible task at the time. How could one ever listen to individuals during migration, short of having a miniaturized recorder on the backs of the birds? But these were the 1980s. It was the time when cassette tape players were the new technology taking over from eight-tracks.

Bill went back to his basement and once again capitalized on his superior mastery of waves. He used the continuous radio wave transmitter he had built for the duck and the rabbit and hooked a little microphone to it. Later, after I had finished my postdoctoral studies at the University of Washington in Seattle in 1998, Bill and I would work together to modify, perfect, and adapt it. But even in those early days, it worked. Bill caught some Swainson's thrushes during their migration stopovers in central Illinois and attached the miniaturized transmitters with their microphones to the birds' backs. It was a neat little package that weighed hundredths of an ounce (no more than 1.5 g).

There was a ton that Bill immediately learned from being able to eavesdrop on the daily lives of these birds. Most importantly, he learned that the thrushes do not only make the loud calls that we hear while they are flying. The birds also whisper to each other on the ground. These whispers are inaudible to us unless we are really close (closer than 16 feet or 5 m). The birds communicate only with each other. The reason biologists had not discovered this quiet form of avian communication before was that the birds keep their beaks closed when they whisper. (Songbirds have a vocal organ, the syrinx, that allows them to do this.) We don't hear any sounds and don't see any signs they are vocalizing, but they are talking to each other. Bill's microphones had opened the doors to a whole new world of animal communication. But Bill wasn't so much interested in the social calls thrushes made to each other on the ground. He wanted to know why and when they called during their migratory flights.

While the task of flying with a bird at night was an unmanageable problem for everybody else, it wasn't for Bill. He went back to his workbench and devised a way to follow individual birds during migration from a car on the ground. He kept continuous contact with the bird using the modified Sputnik radio on its back, but Bill had to keep his radio receiver within 3 miles (5 km) of the bird to receive the signal. Impossible? Not for Bill—but even he was lucky this time around. In the old days, Illinois was known as the thousand-acre farmland. Each farm had 1,000 acres (about 400 ha), a farmhouse, and roads on all sides of the farm. This meant you could travel through the countryside on a grid of farm roads that were only a mile apart. It was so flat that if you drove at regular speed to follow a migrant songbird at night, you might even be able to keep up with it. There may be no other place in the world where you could do this.

Bill's—and later my—task was simple. On good migration nights, which were characterized by midday temperatures above 68°F (20°C) and wind at sunset below 6 miles an hour (10 km/h), we waited until an individual Swainson's thrush took to the air and began to fly. We cut holes in the roofs of our cars and stuck poles through them that we could turn

360 degrees while driving. The receiving antennas were attached to these poles above the car's roof. Bill had an old station wagon; I bought Bill's mother's Oldsmobile for a dollar. We drove like tornado chasers behind a single bird each night, constantly rotating our antennas to determine where the bird was going and to receive the strongest signal possible. All we needed to do was speed after the thrushes while recording their sounds continuously. And that was what we did. What Bill discovered was astounding.

Individual birds had vastly different call rates. Some individuals called every minute on their six-hour flights. Thanks to the microphones attached to the backs of our beloved aerial athletes, we also heard other Swainson's thrushes calling, often in direct response to the calls "our" birds were making. But we also heard veeries, hermit thrushes, and other species responding. And still this knowledge was not enough for Bill. He wanted to know how high the birds were flying and why they were choosing some altitudes and not others.

After the first few birds, when Bill had learned how the birds flew and how to keep up with an individual bird, he would rush ahead to where he guessed the bird he was tracking that night would fly right above his head. And when the bird was directly above him, he would honk the horn of his car. By calculating the time difference between his honking and when he heard the sound in the bird's microphone, he knew exactly how high the bird was in the night sky. This was important because he then realized what it was that caused the bird to change altitude. What the birds were doing was once again very similar to what was happening on the Italian highways of old. A bird would fly up to a certain altitude, call (the bird's equivalent of honking its horn), and check to see if any other birds were around. If there were none, this was probably not a good place to be. So, the bird would go higher up, call again, and see if any other birds responded. If a lot of them did, the incoming bird was probably on the right track. It had likely found an area where there were favorable tailwinds, little turbulence, and a highway headed in the right direction.

What Bill's research told us was how birds on perhaps the biggest aerial migration on Earth were communicating with one another. He had discovered a highway in the sky, where birds were providing each other with key information on how high to fly, where to go, and who to follow. Who knows, perhaps they did not all agree all of the time, but driven by instinct to migrate, once they were out there flying, the migrating birds were learning from their peers. What Bill had discovered was an information highway above our heads in the night sky.

The symphony of migrating birds from the night sky is as beautiful as the symphony of radio waves from the universe. But unlike the symphony created by waves from space, which follows the laws of physics (George's area of specialty), this ancient organic symphony is created by animals as they exchange information across species and continents, and it follows biological laws we have yet to uncover (Bill's passion). The Indigenous people living on the prairies before the Europeans must have been attuned to what these birds had to say. The time has come for everyone else to tune in, as well.

Bill's seemingly simple observation, achieved by giving thrushes little microphones during their migratory flights, changed our ideas about songbird migration. Instead of seeing the birds as genetic machines—mindless automatons following their innate genetic code—we now understood that these birds talk to each other, discussing which altitude to fly at and which direction to take. By communicating with others, each individual bird taps into a communal knowledge bank built up by billions of animals over vast stretches of time. This one study documented the potentially vast stores of knowledge embedded in the cultures of animals—in this case, the migration culture of North American songbirds wintering in Central and South America.

3 A Little Ovenbird Makes Us Think Again

DESPITE ALL THE MONUMENTAL ACHIEVEMENTS by Bill Cochran—foremost among them the technology that enabled us to follow songbirds during their nocturnal migrations through the flatlands of the Midwest—one of the biggest challenges that remained was to understand how these tiny and most delicate creatures managed to cross major bodies of water. How, for example, did a warbler weighing just a third of an ounce (10 g) make its way from Florida to Venezuela or from Texas to Panama? The most important obstacle during migration for the thrushes that Bill and I tracked was the Gulf of Mexico. How did a Swainson's thrush or a wood thrush that wintered close to Barro Colorado Island in the Panama Canal fly nonstop to the shores of Texas, Louisiana, Alabama, or Florida? Another question was weather—how did it affect the birds' decision-making, and how did they predict the weather en route?

Bill and I originally developed plans to put automated telemetry receivers into the FedEx airplane fleet, so that every plane that flew in the Americas would collect data from birds for us. Other plans included a chain of automated receivers along the Gulf Coast and those parts of Central America where birds arrived or left. The receivers would have alerted us when individual songbirds carrying tracking tags embarked on their journeys and when they arrived at their destinations. Since obviously songbirds need to fly nonstop over the ocean (except for the few that stop on boats or oil platforms), these data would have allowed us to calculate their flight speed and direction. We would also have been able to calculate their flight altitude as we knew the winds and could determine exactly which wind at which altitude would provide what kind of support for the birds to arrive at a specific place at a specific time. In the end, Bill and I never developed this system, but a similar telemetry monitoring system was later set up by the excellent Motus science consortium (motus.org).

Some years later, when tracking tags had advanced enough that we could use them on birds to record their position locally and eventually download the data, I traveled to eastern Mexico. As part of our quest to decipher how birds deal with weather, I wanted to see whether frigatebirds could anticipate where a hurricane would make landfall. We got the idea from the local people in hurricane-battered areas, who told us that all the birds were gone before large hurricanes arrived. The people had realized that although in general you can clearly and fairly accurately predict the path of a hurricane from satellite pictures and forecasts, it is still important to know exactly where it will first hit land. Perhaps, Bill and I thought, frigatebirds could help us because they are out on the ocean. They know the winds, the turbulences, the little details of a storm system (such as the temperature difference between the surface of the water and the first layers of air) better than any human-made weather buoy. And the birds probably have their own weather forecast model in their heads.

I traveled to the Yucatán Peninsula, close to Cancún, and from there to the little bird island of Contoy, because at the time the prediction was that hurricanes would now hit this area roughly once a year or possibly even more often. Unfortunately for us—but fortunately for the people who lived there—no hurricane went over or came close for the next five years. But we still believed it was worthwhile to catch and tag some frigatebirds so we could ask them how they perceived an approaching hurricane—should one come along. We had a net we could fire out of a barrel to a distance of maybe 20 feet (6 m) using air pressure, and we used it to catch some frigatebirds sitting in the trees above us. We then outfitted the birds with a snug backpack satellite tag and were generally happy that our project to monitor the behavior of frigatebirds was finally taking off.

Early the next morning we were supposed to tag a few more birds, but our collaborator, Pepe, who was head of the local national park, said he couldn't drive us around in his boat that day to find more birds because he needed his boat to go somewhere. We expected something interesting might be on the horizon and asked, "Well, is there a chance that we could come with you?" He just said, "Hurry up, pack your stuff. Bring your snorkels, masks, and fins." We still didn't know where we were going, but on the two-and-a-half-hour boat ride out toward the open ocean in his super-fast dinghy, Pepe told us, "If we're lucky, we might see some whale sharks." Little did I know it at the time, but although our frigate-bird study was not giving us the results we had hoped for about how birds might predict the weather, what happened next was going to give me a completely different view of how songbirds navigate the open ocean.

Pepe told us that there had been reports of a massive gathering of whale sharks to the south. I had so far only seen these giants—the largest fish in the world—in murky waters in the Galápagos. To see a large gathering in clear water in the open ocean would be a dream come true for any biologist. What seemed like an eternity later, we spotted fins off in the distance. Five, ten, thirty, no, forty (!) whale sharks were hanging out together. Pepe parked his boat to intercept the sharks as they swam slowly through the water, and we jumped into the ocean. We were soon floating in front of whale sharks that drifted toward us like huge starships from another galaxy at the end of the universe. We dove down, trying to swim alongside the sharks, staying underwater as long as we could to enjoy this once-in-a-lifetime spectacle.

Usually when I come up after a dive, my ears are totally plugged, but this time by chance they weren't. As soon as I surfaced, I heard a sound that was familiar, but I couldn't quite pin it down. This sometimes happens when you taste something unexpected in your food. You can't quite place the flavor until somebody says, "Oh, this tastes like wild raspberry." The taste—in this case, the sound—was something I recognized, but what was it? Then I looked up and there it was: a little ovenbird. I had caught this species many times when banding in Illinois and lots of other places. It is one of the most familiar sights and sounds in the bird fauna of North America. I did a double take. An ovenbird out here in the middle of the ocean? It must have been on its way south during its annual migration. I quickly calculated the time it must have taken this bird to fly out here. If it was coming from the Mexican peninsula, it must have left sometime before we did, maybe around sunrise. I assumed it hadn't come from North America because obviously this bird had to fly nonstop over the ocean during its migration, so that starting point would have been much too far away.

At first the ovenbird was just a little speck in the sky. But then it flew down in the typical descending flight a songbird makes when it is about to make a stop during its migration. *This is interesting!* I thought. *This bird is going to land on the roof of our boat.* But that is not what this ovenbird did. Although it could easily have landed on the boat, it landed on the water a few hundred feet (about 100 m) away from me. What was going on here?

My thought was that it must have been one of those birds that didn't have enough fat reserves to fuel its flight; it must have miscalculated the timing of its migration. I was convinced I had just witnessed the death of an exhausted ovenbird on an open-ocean crossing. I couldn't believe I was doing this, but I left the whale sharks behind me and swam as fast as I could in the direction of the ovenbird, so I could grab it before it sank or was eaten by a shark with much bigger teeth than the whale sharks'.

Even though whale sharks look scary, they are harmless because they are filter feeders and their teeth are so tiny you can barely see them. Sharks that use their big sharp teeth to grab on to their prey are another thing altogether. Swimming out toward that ovenbird was one of the scariest moments of my life. I knew what sharks could do because six of them had once circled me in the Galápagos and three had made a serious attempt to attack me. I had to stab their noses with my tent pole, which I always carried to defend myself when I was snorkeling. From other experiences I have had in the ocean, I was well aware that there were creatures a little farther down and out of sight perfectly adapted to their environment that would know exactly where I was as I splashed around making a fool of myself on the surface.

But I took the chance because I really wanted to check out the ovenbird to see what the problem was. The waves were not high for the open ocean, but still fairly high for swimming and for making progress toward the ovenbird. I could only see it when we were both on the top of a wave at the same time. Finally, I was within about 15 feet (5 m). I was happy. I thought I could check how much fat the ovenbird had and ascertain its physical condition so I could understand why this bird had died in the ocean.

I came closer, maybe within 10 feet (3 m). The bird was lying on the surface, wings spread, but its head was out of the water. And then it took off. Seriously. It took off. As far as I knew, songbirds could not land on the ocean. How could this bird sit on the water for five or maybe ten minutes and then take off? But then I saw it had come down again closer to the whale sharks. *Okay*, I thought. *I scared it, and this ovenbird has used up the last of its energy to escape me.* I needed to get closer to find out what was going on. Once again, I felt I was paddling for my life because I sure hated being alone a long way from my group, exposed to sharks in the deep, pitch-black open ocean. After a few minutes, I arrived at the bird. Once again it was lying there, wings spread, body floating, head up. I stopped swimming because I didn't want to scare it again.

I was eerily fascinated. I didn't know anyone who had witnessed natural selection at work in a spectacle of Poseidon against ovenbird like this, but presumably it was something that happened millions, maybe even tens of millions, of times every year during songbird migration. I glanced over to the others, who were having the time of their lives with the whale sharks while I was watching a lone ovenbird die. Suddenly, the ovenbird chirped. It pulled itself out of the water with its wings, gave itself a quick shake to get the water off its belly feathers, and took off. With a few more chirps, it ascended into the sky and continued its migration straight south. I watched until the speck was gone.

I had never been more surprised in my entire career as a biologist. Everything I thought I knew had been thrown overboard by this ovenbird. I'm not claiming that all songbirds land on the ocean. But I know that at least one little ovenbird, south of Mexico, did. For me this meant we needed to question everything we thought we knew about bird migration over natural barriers. It also meant that our models of flight speed, which we based on our assumption that songbirds generally fly nonstop over oceans, were quite possibly wrong. Quite often in science we go by the simplest explanation. The simplest explanation for this ovenbird was that it had dropped onto the ocean exhausted and emaciated and was about to die. But simple explanations may not be the best guide in biology, where every life-form is fighting for its survival and finding innovative solutions that we humans often don't know about, simply because we are not there when the most important events are happening in that animal's life.

Floating there above the deep abyss of the ocean watching the speck that was the ovenbird disappear into the sky, I thought we should give each of these individual animals their own voice. Let each one of them tell its own story and tell it right. And let us not just accept (the way I did) some stupid assumptions that we have made about the way life works on our planet. Perhaps this little ovenbird was actually a hero in its world—a Wonder Woman or a Rocky Balboa of its population—but it convinced me that I would never really know what an animal was doing unless I had a tracking device on it.

4 The Early Days of Tracking

TRACKING MIGRANT SONGBIRDS in Illinois with Bill in the late 1990s was an amazing experience, and it inspired me to study animal interactions within and between species in the wild. At the time, however, it was extremely frustrating. The birds would disappear after Bill and I had tracked them for a day or two. We would never know where they came from, what other birds they interacted with, how the time we observed them fitted in with the rest of their lives, and whether they survived that migration or died.

The entire tracking enterprise was also challenging on a personal level. To be able to be ready any time of the day or night during the six-to-eight-week migration season of these birds in spring and fall, we had to cancel all travel arrangements as well as all social commitments. It was impossible to accept a dinner invitation because I never knew if I would be able to show up. And even on the days after a chase, I could not be sure what I would be able to do because we followed these birds from roughly half an hour after sunset to about half an hour before sunrise. Driving continuously behind them, trying to speed up when we needed gas so we could fill up the tank in the shortest possible time, then rushing up behind the beeping signals from the night sky once again while simultaneously avoiding speeding tickets, was exhausting.

Occasionally we had to drive a wee bit above the speed limit so as not to lose the birds, which more than once resulted in having to stop and talk to an interested traffic cop who wanted to see why there was a huge antenna sticking out of an old car, an antenna that was constantly changing direction as we traveled with the bird. Then our delicate task was to try to indicate that although we would love to explain exactly what we were doing, we didn't have time to talk right then, because otherwise the little songbird we were tracking in the night sky would forever disappear into nothingness. We tried to be as serious as possible, because a few times we could see that the cop did not believe us, but if we lost the bird, which might happen if we talked for more than a couple of minutes, we would miss all the information about where this bird was heading and how it was going to react to its new surroundings the next morning when it landed, and we would not be able to show these data to the rest of the scientific world.

But my friend Roland Kays and I, in consultation with Bill, felt there was a solution to our problem. For his PhD in Panama, Roland had worked on kinkajous, arboreal raccoon relatives that turned vegetarian. He had tromped behind these canopy acrobats on foot through hundreds of jungle nights to track their movements, but despite his Herculean efforts, he had gotten only a small handful of data points. Together, we wanted to change radio telemetry forever and for the better to spare future researchers such long and fruitless night treks. A year or two later, when I started my faculty position at Princeton University in 2000, Roland and I teamed up to study species interactions in the wild.

We told ourselves that we were clever and adventurous and forward-looking. We decided we would find a little universe mostly separated from the rest of the world and study animal interactions there. What could be better than an isolated island? It was the time when Rosemary and Peter Grant, friends of mine at Princeton, had already started to show that evolution can happen in real time on isolated islands such as Daphne Major, the tiny volcanic spot in the Galápagos they had chosen as their study area. But we wanted to go to a place where a lot more interactions happen than on a desert island in the middle of the Pacific.

When I was doing my postdoctoral research in Seattle in the mid-1990s, I had worked at the Smithsonian Tropical Research Institute on Barro Colorado. That small island paradise in the Panama Canal was where Roland and I first met. Now we both decided that Barro Colorado would be a perfect research site. It was almost 6 square miles (15 km²)—large enough for many species to live there, but small enough that we thought we could track the animals continuously using radio telemetry so we could study their interactions.

This was where Bill came in to help us design our system. All we needed to do was build seven towers taller than the trees in the rainforest to receive radio signals from the tags we were going to put on the animals. Then we were going to triangulate the radio waves from the tags to find out the animals' locations and use some tricks from Bill's and George's playbooks to tell us what the animals were doing. The principle of radio engineering behind our automated radio telemetry system, or ARTS, was quite simple. When the radio telemetry tag on an animal emitted a signal, one tower would be to the left of the animal and another tower would be to the right and perhaps farther away. As a result, the radio signal from the animal would arrive at the two towers at different times and from different angles. At the same time, the radio signal would be wobbly if the animal was moving and beautifully clear if the animal was stationary. At least that was our plan.

But building a telemetry receiver system in the rainforest turned out to be a bit of a nightmare. With the island being so close to both oceans, the Pacific and the Atlantic, there was not only water in the air but also a lot of salt, which quickly corroded metal and ate the antenna wires. But at this time in history, radio telemetry was our only choice. Using GPS (the Global Positioning System) was impossible as there was no GPS reception in the forest yet. And so, the only way to get the position and behavior of an animal in its environment remotely was by determining the direction of its radio signal from different towers. It was a major undertaking and also one where after a few years you think, *I would never do that again.* But at the time, we needed all these practical and data developments to sort through our thoughts so we could prepare for an even bigger adventure.

What worked was that we could attach radio telemetry tags to many different species. There were lots of biologists out there who knew their animals so well that they could tell us where they were and collaborate with us on the fundamental problem of how best to attach an electronic tag to a certain animal. After a while, we got a good understanding of how wearables for wildlife, as we started to call them, could work and how to attach tags so they had a minimal impact on the animals' natural behavior. Tags had already been tested in the global animal research community, but our research team at Princeton University, the Smithsonian Tropical Research Institute, and the small global community of biologists working on wildlife telemetry developed many new methods to attach smaller and smaller tags with fewer and fewer issues for the animals. Toward the end of the project, after an encounter with migrating dragonflies on Cape May that I describe in the next chapter, we even designed tags tiny enough to be attached to insects.

Our community of animal ecologists was excited. Using the new improved tags, we could study monkeys, coatis, agoutis, ocelots, woodpeckers, white hawks, sloths, and even orchid bees. Many students and colleagues from all over the world joined us at Barro Colorado Island during the handful of years when our telemetry system was fully operational. At this time, we thought that a 2-ounce (50 g) radio collar for an agouti was actually lightweight and ergonomic. Little did we know that one and a half decades later a much more elaborate tracking device, almost as powerful as a computer was during the ARTS project, would weigh not even a quarter ounce (just 5 g). It could be fitted to the ears of a small mammal as a true wearable, a possibility we had not even dreamed of in the days of our ARTS project.

The computer in the central laboratory on the island was next to the cafeteria, and we and everybody else could always look to see where the animals were in real time. In the wee hours of the morning, when we wanted to go out to observe our animals at sunrise, we could find little colored dots on the map telling us roughly where to find them. This allowed us to head out in the right direction, sparing us an hours-long or day-long—and previously often fruitless—search for our animals in the forest.

Having real-time information on where animals were so we could go out into the forest to find them and spend long hours observing them directly was one of the most exciting and helpful aspects of the system. As we watched the dots in the lab, it became increasingly clear that individual animals were constantly interacting. Agoutis spend their lives in fear of ocelots, and we now realized why. Ocelots were searching for them all the time, following them and ambushing them. The information on these interactions allowed us to map out not only the agoutis' "landscape of fear"—the most dangerous places for them to be—but also their "landscape of energy expenditure." How much energy were individual agoutis expending in different locations and how did they move between locations? How often did they move? Where were their well-trodden trails, and did they establish new trails or use a network of ancient and culturally inherited trails? Did some individuals really cover an entire area by crisscrossing it? We could draw similar maps for the arboreal species we studied. What kind of highways were the monkeys using? Or the kinkajous, those funny raccoon relatives that eat fruit and drink nectar in the canopy and only move at night?

What we were learning every day highlighted moments from these animals' lives that were completely new to us. We could finally study who was eating whom, and where and when. We could immediately see when an agouti was killed by an ocelot, or when an agouti was carrying away a nut that had fallen to the ground under the mother tree. Most often, the agoutis moved the nut to a different location and then buried it. We realized that there was a conflict of interest between the tree and the agouti. Trees with big seeds, like the black palm (chumba wumba), need big rodents like agoutis to distribute them. However, the tree is only interested in the agouti carrying the seed if the agouti moves it away from the seed shadow of the tree: most seeds that fall within the seed shadow will not sprout because local competition is too high, diseases and fungi may destroy seeds, and it is often dark. When the agouti buries the seed, it is putting up future food reserves in a process called hoarding. What interests the tree most, however, is the presence of ocelots. Ideally, the agouti will be eaten by an ocelot so that the seed the agouti has carried away from the parent tree will remain uneaten and can sprout and grow into a new tree.

Many times the signal of an agouti flatlined, often in the middle of the night. When this happened, we were woken up to assess the situation, and usually somebody had to run out to install a camera trap to see what had killed the agouti. Using camera traps, we were able to remotely detect exactly what happened when the killer returned to the scene of the crime to consume its prey.

Our automated radio telemetry system, combined with lots of on-the-ground field observations, also allowed us to study the movement of the orchid bees that are essential, as are many insects around the world, for transporting pollen between plants. To our great surprise, these supposed homebodies traveled long distances. They even went from the island to the mainland to find new trees to pollinate, and then returned to the island a few hours later. What an incredible effort by the bees, and what superb manipulation by the trees—using scent distributed over long distances on the wind and synchronous flowering times—to make the bees fly huge distances to ensure the trees' survival and propagation.

These were exciting times. Every day was another expedition into the unknown to explore new behaviors and figure out new scientific directions. But let me talk about the failures, as they were actually much more important in the long run.

ARTS was an amazing technological feat, but with all the projects and the (at least partial) successes, it was becoming difficult to organize all the data, especially all the metadata. Metadata are the data you need to really understand your results. Metadata are things like which animal was stalked by which other animal, which tag was used, which radio frequency the tag had, and how the radio frequency changed with temperature, humidity, and animal activity. Did the tag have a mortality sensor or not? What was the output power of the tag? How long did it last? What attachment technique was used? All this information had to be organized in a way we could use it, and at the time we didn't have a good system for that. In addition, the data from the telemetry towers streamed continuously to the central laboratory, where we saved everything in folders but not systematically in a database. All these data, particularly all the information from the animals that needed to be interpreted, turned out to be overwhelming.

Field ecology at the time did not have a way of organizing data on a large scale because in most studies it was not needed. Most field ecologists studied one or very few species, at one general location, and the data from the tags and animals were few and far between. Bear or wolf researchers, at least in the early days, concentrated just on their species, and not on the interactions between predator and prey and scavengers such as ravens. Marine mammal researchers would often study the locally abundant species within a large colony. We, however, had data arriving every second from a large number of animals and species. We needed to be able to split up the continuous data stream into various animal information systems in different studies, and the data needed to be farmed out to different researchers in as close to real time as possible. You could think of it as a system where you have to keep track of cell phones. The cell phone companies need to know each account and each person's utilization of the network. The companies also need to keep track of incoming and outgoing calls and charge each individual for what they use. This is when we came up with the idea for Movebank (movebank.org).

Movebank was created to preserve and archive the movement and rhythm of life across the planet. Consider a museum, which is basically what existed at the time. The exhibits were static. They were both interesting and boring because they told us about the animals, but the animals were missing their own detailed stories. Our knowledge of animals was just the outer shell of a living being. What we wanted to achieve with Movebank was a history of the living pulse of the planet.

Imagine what information we might have today had Movebank existed before European invaders spread west through the North American continent. We would know which animals the Indigenous people observed, and what tools and approaches they used. We would have a glimpse into the seasonal movements of the bison on the Great Plains. We would know how the wolves and bears interacted with them, and in turn how both wolves and bears roamed the Greater Yellowstone Ecosystem and beyond. How did Lewis and Clark interact with the local deer? When and where were the salmon running that fed the expedition through the winter? Later in the same century, where did the passenger pigeons roam and feed? Thomas Jefferson and German explorer Alexander von Humboldt would have been able to discuss how oilbirds moved in the Amazon rainforest based on the Movebank records that Humboldt and his fellow traveler Aimé Bonpland would have entered during their expedition in Venezuela. (But more on that later.)

One night on Barro Colorado, we sat down with a glass of good Panamanian rum and ruminated over our data failures. We needed to scale up our approach to be ready not only for tens but for hundreds or even thousands of species. At the same time, we needed to find a way to organize and archive the data so they would be available forever. We also needed to provide tools for researchers and the public to interact with, and learn from, any animals they might want to tag and track. In principle, what was needed was an Internet of Animals where information was stored, shared, and archived. At the time, we didn't know where this idea would lead, but we knew this was where we needed to go.

As we were grappling with all this, we realized there was an even more important flaw in our system that affected everything we were doing. Our data—well, our animals sending us the data—were telling us that our entire approach was wrong. We were learning the hard way that not only were the animals interacting constantly, but they were also not staying in one place. That is, they were not staying in "our" rainforest plot, even though everybody had told us this was for the most part a completely isolated island. The bats were coming and going from the island. The orchid bees were flying off and coming back. The coatis were often swimming across the canal. And even the monkeys were taking trips to the mainland. But what really astonished us was that our ocelots were also coming and going, and some of them paid a high price for their urge to travel.

One of the ocelot collars started to beep from a lagoon on the edge of the island. When we went to the location, we realized that the radio signals were coming from somewhere under the water. Luckily the water was not very deep and we were able to locate the collar at the bottom of Lago Gatún. It was clear immediately when we retrieved it that something bad had happened to this poor ocelot. We found the bite marks of a crocodile on the collar. Apparently, the ocelot had been about to leave the island when a crocodile caught it, bit its head off, removed the collar, and then swallowed the rest of the poor animal. Or—even more gruesome—it might have swallowed the entire ocelot and then spit out the collar. A terrifying thought, but one that made us realize that our supposedly isolated world was not isolated enough.

The little self-contained universe that we had wanted to study turned out to be one node in a big travel and communication network. In retrospect, this seems so obvious. How could we ever have thought that a seemingly isolated island was not also highly connected, especially with all the biodiversity in a larger rainforest environment? There are some species that do not cross the water and spend all their lives in one place, but even in those species, there are always individuals, the Polynesians of the animal world, that leave their homeland, their beloved little island or shire, and go exploring to find greener pastures or adventure. It turned out to be mostly the juveniles that felt the urge to leave the island, unbeknownst to many of the researchers there. This revelation was key to a successful global system to come. While we initially thought we had failed miserably, we probably learned the biggest and most important lesson we could ever learn: the world is an interconnected place and animals are interacting beyond any borders, physical or political, that we project onto them.

In honor of our failure, we called ARTS our Fitzcarraldo expedition—a monumental undertaking that ultimately reached its goal, but wasn't sustainable. Fitzcarraldo, in Werner Herzog's movie of that name, was a famous fictionalized rubber baron in the Amazon who was trying to build an opera house in Iquitos and get Caruso to sing there because he was a major fan of the Italian tenor. Fitzcarraldo was as foolish as we were at the start of our rainforest adventure. We thought we were going to solve a major issue, but then we realized that no matter how much effort we put into it, we were never going to solve our problem because there was a systemic flaw in our original assumption. Our failure was that we thought we could study nature's principles of species interactions on an isolated island. Fitzcarraldo's failure was the assumption that he could make enough money from his wild rainforest expedition to build an opera house in Iquitos. And yet, despite his failure, Fitzcarraldo brought Caruso and the entire opera cast to Iquitos for an open-air performance. Both Fitzcarraldo's and our grand experiments succeeded in the short term, but failed in the long term. And each probably told the dreamers more about the world at large than they had anticipated.

The general lesson for all of us is to explore, to fail, and to discover why our assumptions are wrong. Then regroup and get going again. Life is simply too exciting not to get out there and see what you can do.

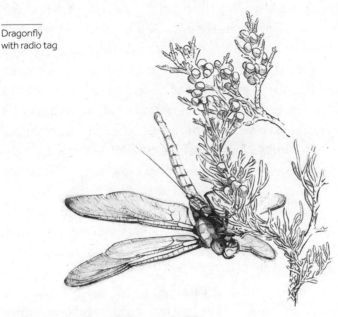

Dragonfly
with radio tag

5 Walking Like a Cowboy

THROUGHOUT MY CAREER, I have gone out and done field-work to test my ideas and come up with new ones. Fieldwork around the world is not without its perils. You can prepare for so many things—and I think of myself as somebody who is extremely well prepared—but there are also a lot of things you just can't prepare for. I'd like to tell you about one now. It is also the origin story for those tiny transmitters we used for orchid bees in Panama.

In 2005 I was an assistant professor of ecology at Princeton University. My friend and colleague David Wilcove, a conservation biologist, and I went birding in Cape May, New

Jersey. We walked along the beach trying to spot rare shore-birds. It was an extremely hot day and wherever we were the shorebirds weren't; however, the sky was full of specks. I did not focus on the fuzzy stuff whooshing across my binoculars because I wanted to see birds, and so I continued to scan the beach. After a while, I heard David say, "I think this is migration day for dragonflies." I lowered my binoculars and focused on the fuzzy stuff flying back and forth instead—and what I saw were hundreds, thousands, tens of thousands of dragonflies. I later learned that some were flying up and down the coast and others were flying along the coast in one continuous line southward. It was, indeed, a massive migration. Thankfully, David teased me and jokingly said, "You'll never track those."

That same evening, I called Bill and told him that we needed small, super-lightweight transmitters for dragonflies. They should weigh a maximum of nine-thousandths of an ounce (250 mg). As always, Bill said, "Oh yeah, I thought about that about ten years ago. It wasn't possible back then, but I think the crystals are small enough now. Let me call Jim." Bill's son, Jim Cochran, had already gotten a hold of some of the small crystals without telling us, and he already knew how to miniaturize the transmitter circuit boards. You can't imagine how surprised David was when, two weeks later, we had our first nano radio transmitters, small enough to fit a dragonfly.

We tuned the transmitters to the 400 megahertz range so we could reduce the length of the transmitting antenna to about 1½ inches (4 cm). That was much shorter than the 6-inch (15 cm) antennas we used on our regular transmitters back then, which were tuned to the 150 megahertz range. We were ready to track migrating dragonflies.

The only problem was that we could not catch enough dragonflies. We had thought this would be the easiest part of our job. Together with our colleagues Mike May and David Moskowitz, David and I stumbled around like fools in a beautiful meadow close to Rutgers University in central New Jersey. Each of us was armed with a large insect net on a long pole. We were jumping up and down and sideways trying to catch dragonflies, to no avail. They were just too smart and too quick for us. The next morning, we had better luck. We went out when the air was still cold and now we warm-blooded mammals had the advantage over the cold-blooded insects, which could not react as fast in cold air. After a lot of effort, we managed to catch fifteen dragonflies out of the thousands that were passing by on their migration. We carefully attached our new nano tags—and there we were, the first humans that could follow individual insects during their migration.

I rented a small Cessna airplane from the Princeton airport and flew up and down New Jersey and neighboring Pennsylvania and Delaware tracking the little fliers. After a while, I figured out how to locate them to within about 500 feet (150 m) on the ground. Now we could combine aerial and terrestrial searching. I tracked the first of our migrants down from the air to the edge of a forest near Cape May. After I landed the Cessna at the Cape May airport, my team picked me up and we continued the search for our little insect friend on foot. We located the signal in an eastern red cedar tree. We stood around the tree, five pairs of the world's finest binoculars scanning the tree from root to crown, but we could not find the little critter even though I knew from the radio signal that it had to be there.

I decided to climb the tree to get closer. I really wanted to see with my own eyes how a dragonfly spends its time during a stopover. Unfortunately, years ago in my early years studying biology in Munich, Germany, I had twice failed my plant identification exam. I am simply bad at recognizing plants. And yet, I probably should have learned and known that there is poison ivy in the eastern U.S., even though I had myself migrated over from Europe. Standing in front of the cedar where the migrant dragonfly was hanging out, I had thought that was plain old ivy growing up the tree. It was summer. I was wearing shorts. And the tree was hard to climb. I had to hug the trunk with my bare thighs to slowly crawl up the tree, and later slide down it.

I was ecstatic when I found the dragonfly. To be more precise, it was a green darner. In German, it is known as the American king dragonfly, which it truly is, a venerable trooper in the royal insect air battalion. It was hanging from and blending in with one of the top branches, completely calm and stationary, with its translucent wings hidden by a backdrop of foliage so no predator would be able to find it. I was so proud. It was not only the first insect tracked during migration via a signal from a radio transmitter, but I was seeing it with my own eyes as it rested in that tree at the beach.

After I slid down the tree trunk, I pointed out to everybody where the dragonfly was so we could all see it from the ground with our binoculars. I couldn't have been happier about our discovery. It was helping us understand the dragonflies' behavior during their southward migration, and now we could even study their stopover behavior. What I learned only a few hours later was that poison ivy really burns.

The next two weeks, tracking dragonflies from the plane, I had to have ice bags between my legs and I constantly used ice-cold yogurt and cucumber slices (one of my grandma's remedies for burns) as well as medical cream to ease the pain. When walking, I had to keep my legs as far apart as I could. I finally understood why cowboys walk the way they do. Back at the university, I didn't tell people what happened. I just walked carefully and slowly because you can't tell your colleagues or your students that walking is painful because you seriously burned your inner thighs with poison ivy. In retrospect, I guess I could have been better prepared had I studied harder for that plant identification exam.

6 Our Sputnik Moment

IN FEBRUARY 2001, my team and I were by the Panama Canal in a little village called Gamboa. I was teaching part of the Princeton University field course in ecology and we were conducting research in the Panamanian rainforest. It was only a short boat ride from the island of Barro Colorado, where we had just started to build our ARTS (automated radio telemetry system).

George Swenson, my old radio astronomy friend from Illinois, had arrived the previous day. We were going out birdwatching in the rainforest on the mainland, just across from Barro Colorado. George had been an outdoor enthusiast in his youth, even climbing Denali in Alaska. But now, at almost eighty, he was no longer doing so well walking. This was why we were taking one of the Smithsonian Tropical Research Institute boats, a small dinghy with an outboard motor, so we could scoot into the beautiful lagoons in the vicinity of "our" island. Our surroundings reminded us about the times when the German naturalist Alexander von Humboldt explored what he called "the Equinoctial Regions of America" in 1799, five years before he visited the third president of the newly formed United States of America, Thomas Jefferson, for three weeks in 1804.

It was an amazing morning. We saw a harpy eagle, the largest and most majestic raptor in Central America. And we were lucky enough to see a swarm of army ants. These raiding hordes are true miracles of nature. Millions of fearless little colony members chase all the insects and spiders and scorpions and froglets up from the forest floor. Everything that can move will move when army ants are in the neighborhood. And when army ants are about, there is danger in the air too. Army ants are always accompanied by large flocks of different species of antbirds, some of which forage exclusively around these ants. These birds have thrown their lot in with the ants. If the ants don't raid, the birds will not survive. These birds try to pick up all the food—insects, spiders, and more—chased up by the raiders. The poor animals on the jungle floor have to choose between a rock and a hard place. Either they stay on the ground and hope that the ants will miss them—which is unlikely, because army ants on the march are as thorough as a hurricane or twister coming through—or they jump up, fly up, or scurry up onto trees and branches, hoping they can escape the antbirds. This is also a long shot as antbirds are aerial acrobats of the highest order.

They would beat *Top Gun* pilots hands down—or should I say wings up—because they have spent their lives targeting fleeing insects and spiders.

A swarm of army ants is an incredible spectacle. It is local mayhem, horror on six legs. Millions of ants collaborate to kill and butcher everything that gets in their way. These ants spend their life on the move. They bivouac and build a new mobile home constructed from living ants every night. The queen is kept deep inside the collective body, guarded by soldiers and other workers that build a complex three-dimensional support structure around her using nothing but their own bodies. These ants, individually, don't have much of a brain, but as a collective, they can build bridges and run back with food from the raiding front along self-organized highways. That way, they optimize the search for food so that even though hundreds of thousands of ants are moving through the forest in different directions, they don't get snarled up in traffic jams.

While I was in Panama, my team and I wanted to take a closer look at the interspecies relationships between antbirds and army ants. We wanted to understand whether army ants and antbirds really do benefit from each other, or whether antbirds exploit the little knights of the forest by stealing food the birds would otherwise gather on their own. Do the birds really need the ants or are they just being lazy? We hypothesized that both species benefit if the air force keeps the insects grounded so that the infantry can raid them all. The aerial forces would eat only those insects that would fly off anyway.

We started by deciding how we could test our hypothesis in a way we could measure. We came up with a simple

experiment. We would count the food items being transported back along the ant highways toward the queen safely
tucked away in the raiders' bivouac. Then we, the humans,
would appear and chase away the antbirds for a short while.
If ants and antbirds benefit from each other, there should
be less food transported back to the queen during this time.
If the antbirds steal food that would otherwise belong to
the fearsome knights, the ant cargo group should transport
back more food during the time we keep the antbirds away.

How can you scare off the antbirds without disturbing
the ants? Simple, we thought. We would position enough
students within the raiding hordes so that once the ants had
surrounded them, the students could shoot in the direction of
the birds with water guns to temporarily scare the birds away.

As usual, the animals taught us a lesson. The antbirds had
already determined that we were not a major danger. We had
been around too long to be real predators. Antbirds of one
species, especially—the ground cuckoo, a majestic badass of
the rainforest understory—decided that they were the boss
and did not care at all about our attempts to scare them off.
Imagine a phalanx of highly motivated students, shooting
water guns in the forest at tiny antbirds to scare them away—
with the birds completely ignoring them.

In the end we managed to create a few short intervals
when fewer birds were around and were able to suggest that
the ant-antbird interaction was a parasitic one, meaning the
antbirds might be the winners in this interaction, stealing
food the army ants would otherwise get for themselves. But
imagine the faces of the birders listening to our story that
night. Panama, particularly the Pipeline Road area, is *the*
spot in the world for birders to search for the highly elusive

ground cuckoo. It's one of the best and most desired sightings when it comes to neotropical avifauna, and every birder who visits here hopes to check it off their list. We had interacted directly with the most elusive bird they wanted to get a glimpse of from afar through the rainforest understory. The birders looked as if they wanted to jump at our throats and do nasty things...

The army ant–antbird connection is a reminder of how much we can learn from nature. Not only from the morphology of animals and plants—things such as the self-cleaning properties of lotus leaves, used in paints for protecting walls or as self-cleaning glass in traffic control sensors, or the adhesive properties of gecko feet, which now help robots climb walls—but also, and more importantly, from the behavior of animals. The behavioral patterns they have evolved and the rapid adaptations they are capable of making will guide us toward a better future for humankind. I assume we can all agree that optimizing our own well-being across the entire planet is a major challenge for us as a species. The goal as I see it is to move from mechanical biomimicry to behavioral biomimicry—and we clearly have much to learn.

Humans could benefit in more ways than one from lessons learned from army ants. The ants collectively adjust their behavior to safeguard their future because they can only survive as members of a group. They also need to safeguard the rest of their environment because they can only survive if their environment is intact. They can't overharvest in one spot as this would kill everything that lives there and turn the place into a desert. You see what I'm aiming at here. The tragedy of the commons is as much an issue for the tiniest of our fellow beings as it is for us, and even tiny creatures that seemingly brainlessly raid their environment have much to teach us.

But back to our birdwatching trip with George. The next morning, we witnessed the migration of two major groups of animals. Vultures were crossing the Isthmus of Panama on their way back north to spread across most of North America. They were flying low because it was early; the thermals had not yet started and the winds higher up were not in their favor. Quite often we miss the massive migrations of hundreds of thousands of individual hawks, vultures, or other raptors because they fly so high they are invisible to the naked eye. It is only when you look through your binoculars that you see the sky is filled with an infinite number of dots all moving in the same direction.

The other migration we saw was of Urania moths. These moths are daytime fliers. They migrate southeastward across the Panamanian rainforest and go—who knows where. Perhaps Venezuela, perhaps Colombia. Nobody knows. We were watching the movement of hundreds of thousands or millions, perhaps tens of millions, of individuals that had apparently had a really good season over in the west, toward the North American continent, maybe in Costa Rica or even Nicaragua. This happens every few years, and the moths then migrate out of their established range into new areas. Or maybe they migrate every year, and we just don't see them. Maybe they only migrate over Lago Gatún, where we were, in years when the winds are perfect. Perhaps they sometimes fly in less densely packed groups because of winds. Maybe some years the migration is more spread out over time. Or maybe when the winds are unfavorable, they stay closer to the ground or fly within the rainforest.

Or maybe they sometimes migrate at night. There are more questions than answers, and we don't even know what we should be looking for. This is typical of our ignorance of migration systems around the world: we know next to nothing. My mind flew with the Urania moths overhead until they disappeared.

In the afternoon we took two trucks and drove across the Isthmus of Panama to find good places for future studies of hawk migration. At one point we turned down a dirt road and drove along because there was no sign to stop us. At the end of the road, we found a swampy area filled with tallish grass and surrounded by trees. We saw some broken-down buildings and what appeared to be telegraph poles with wires hung between them.

At first, we had no idea what this old structure was. Then suddenly George got very excited. After World War II, George had worked at what would later become the NSA, the National Security Agency. He was one of the top radio engineers at the agency and had advised on many aspects of radio communication, among them the optimization of what was known as the Wullenweber array system.

Very few people have probably heard of the Wullenweber array system. It was a top-secret project consisting of eighteen listening stations around the world. Like ARTS, it tracked radio signals, but the radio signals it was designed to track were coming from certain ships in the ocean, mostly Soviet submarines. Submarines have to come to the surface to communicate with their base, which, in this case, was Moscow. During those few moments when a submarine surfaced and communicated, all the Wullenweber arrays around the world were listening and trying to triangulate the position of the submarine. In a way, it was exactly what we were doing in the rainforest, except we were trying to locate the position of an agouti, an orchid bee, or a palm nut.

George was ecstatic. He had not followed the fate of the Wullenweber arrays after they were decommissioned in the 1980s, and he was not aware that one of his beloved systems had been built in a swamp right next to the Panama Canal. But here we were, tromping around the old antenna shop and swinging around the old telegraph poles on the metal cables that once connected them and provided the link to Soviet submarines that had just surfaced anywhere in the ocean.

The part of George's story that stuck with me was that the radio waves from these submarines bounced when they reached a certain level in lower space. This allowed them to travel huge distances to their destination, which would be a receiver somewhere in Moscow or other Soviet territory. Before this, I had not given much thought to how far relatively weak signals could travel thanks to the qualities of the boundary layer between Earth's atmosphere and the vacuum of space. I tucked this information away just in case I found a use for it later.

Most evenings after our excursions, we sat outside on the steps in front of the old houses in Gamboa, which had been built about one hundred years earlier during construction of the Panama Canal. The houses are on stilts to allow air to circulate beneath them and through them to keep them cool on hot summer days. There was no air conditioning a hundred years ago, and even today many houses still don't have air conditioning. This passive cooling system is an easy way to ensure comfortable temperatures inside. But after our long days in the field, we had taken to sitting quietly outside as the sun set, each of us with a good drink in our hand. We watched the parrots fly by. We didn't know where they spent the months when they were not in this area or where they roosted when they were here, but at least it was exciting to see that they usually flew in pairs and all met up somewhere in the trees behind us. With all the new information we had gathered over the past few weeks from our animal tracking data and the amazing glimpses we had gotten of natural systems in action, we were mulling over what it all meant to us and our future research efforts.

George was the first to organize his ideas. "Your ARTS system is exciting and allows you to understand local interactions, but like the Wullenweber array, it will not have a long history. You have to think globally for one, and you have to think about the interactions and movements of all the players, not just one species, like the submarines. You need to think bigger." He paused. "You ecologists have a huge responsibility to the world and you're not living up to it. You think too small, you don't organize yourselves globally, and you don't demand the instrumentation you really need to answer the questions governments and society at large ask— or should be asking."

Wow. This kind of pithy statement was pretty typical of George. As someone who had installed one system to study the universe and another to listen for messages from outer space, and as someone who had designed these systems to last for hundreds of years and over multiple generations, he tended to see the big picture. I was half offended, half dumbstruck—and getting more and more convinced George was right.

How could we ever make progress in animal ecology around the world if we didn't know the fate of an individual from its birth to its death? If we didn't understand its decisions, or where it wanted to live, or what habitat it needed, or what interspecies connections it had? We had to know all this to know what we needed to do to protect it. And, if we didn't understand its lifestyle, then we couldn't learn from its behavior and mimic what is needed to safeguard ourselves in the future. If we wanted to follow up on the idea of behavioral biomimicry, which is likely to become something big, we really needed to understand the individual and collective decisions of animals around the world and understand how we as humans could adopt some of their reliable and proven principles into our own decision-making processes.

I replied in what I hoped was a measured and tactful manner. "George, I think there's some truth to what you are proposing. You mean, we should go global and study many species? This is a huge endeavor, but…" George wasted no time coming right back at me. "Yes, it is. But we had the same challenge forty years ago when we started bringing together radio telescopes across the U.S. and around the world. Every university, every outfit had their own telescope. Each individual telescope saw almost nothing, but if you link them together, which is what we did in the Very Large Array, that gave us an almost perfect view of the universe. Linking them was the only thing that made sense."

Our thoughts were interrupted by the massively deep honking of a horn from one of the huge tankers navigating the canal. Perhaps this stentorian blast was a reminder that you can implement big thoughts if enough people think they are important. About a hundred years ago, somebody got the idea that we needed the Panama Canal. It was a bold idea. The first attempt failed miserably. Not because of civil engineering, but because people did not include nature as one of the most powerful forces in their decision-making; they only included physics. But workers on the canal dropped like flies from yellow fever, malaria, and other diseases. The solution was to try to understand where malaria came from, how it was transmitted, and what could be done to prevent it from killing all the workers. Only when the study of nature was included in the planning process was it possible to keep the workers healthy so they could continue the excavations. Learning about nature was what turned the larger-than-life idea of a canal through a tropical isthmus into the reality before us now.

That day, sitting by the Panama Canal, I reflected on the significance of the messages we can learn from nature and, even more importantly, on the idea that if we work together using this information, everyone benefits.

"How would you implement this, George?" I asked. George was quiet for a while because he was deep in thought— although it could also have been because the cicadas had started up their massive orchestra at the same time as the parrots were flying by, shrieking. Then George responded. "In my early days with the Navy after Sputnik was launched, we shot up some experimental satellites using rockets. It's amazing how well communication works from ground to space because there's nothing to get in the way of the radio waves. All the other natural systems, like plants, geology, and many more, are already being studied via satellite around the globe. I think you should set up a scientific system designed to study animal life on the planet using satellites."

This was perhaps the point at which to call the practical guy. I picked up my phone and gave Bill a call in Illinois.

"Bill, we're sitting here at the canal and we're thinking that in addition to having local receiving towers for radio signals, we should also have a bunch of satellites receiving the signals. What do you think?" And instead of being shocked or bewildered that we were thinking beyond his ARTS, Bill simply said, "Let me calculate that. If I take the radio power of a little animal transmitter and the frequency on which we are sending our signals, I can easily multiply that by distance and that would get us into low Earth orbit. So, the International Space Station would actually be one place where this could be implemented. If I could call the astronauts now, give them an antenna and one of our wildlife receivers, they could listen to the tags of our animals on Barro Colorado. Yes, I think it could be done."

I turned to George. "Well, George, this is such a beautiful idea and so simple. How long do you think we need to implement it?"

And George said: "Well, we needed fifteen years for the Very Large Array. But there were already major organizations in place. You will likely need longer."

I laughed. "George, you're getting old... This is such a beautiful and powerful idea. I think we can get this going in four years."

It is 2023 as I write this; it was 2001 back then. George would be proven right and I horribly wrong—by a factor of five.

Galápagos rice rats

7 Rats! Still So Much to Learn

AFTER THE PANAMA TRIP, I had ample time to think about what I already knew about animal behavior and where I wanted to know more. I was fascinated by the idea that they were observing us just as much as we were observing them. I thought about some of the other research projects I had been involved with, and it finally dawned on me how important this insight was.

For most people, rats are probably their ultimate nemesis even when they've never seen a rat in the wild. Rats' reputation precedes them and mere mention of these animals conjures up images of dirt and disease. Rats—except maybe for the courageous rat in the movie *Ratatouille*—have a horrible public image. But rats are actually my favorite animals. Not any rats, but specifically the rice rats on Isla Santa Fé in the Galápagos.

These rats are truly special. There are probably fewer people who have seen the rice rats on Isla Santa Fé than have been to the top of Mount Everest. Galápagos rice rats only exist on this island of 9 square miles (24 km²) in the middle of nowhere. These beautiful and entertaining creatures are descendants of rice rats that arrived in the Galápagos from South America long before humans even knew this archipelago existed. The ancestors of today's Galápagos rice rats must have floated here on huge rafts of vegetation, the same way the tortoises and the famous marine iguanas got here. Together, or separately, all these animals landed in this new-to-them world when their huge patches of vegetation— broken off from forests in the coastal area of Guayaquil in the lowlands of Ecuador—drifted ashore after a voyage of hundreds of miles over the open ocean. Reptiles are resilient and seem to do okay during long ocean trips, but very few mammals can manage such a challenging voyage. However, I never wondered how come my little rice rats made it safely. They are so smart and innovative I suspected they might even have helped the tortoises survive.

Before we had our field camp on Isla Genovesa, where I listened to Baby Caruso's beautiful songs, I worked for years on Isla Santa Fé in the central part of the Galápagos archipelago. At the time, at least two members of the four-person research team were biologists from Ecuador. Our local colleagues— and later friends—hated rats. Before making it to university, some of them had grown up in the slums of Guayaquil, an area riddled with rats. In what was an almost instinctive reaction, when they saw a rat, they would always throw a rock at it. I count it as one of my biggest achievements in my career as an educator that I convinced many of my Ecuadorian colleagues not to throw rocks at the rice rats of Santa Fé. I convinced them that these rice rats were almost like kangaroo rats or elephant shrews. Conjure up an image of whatever you think is the cutest, most beautiful little mammal in the world—and then imagine a Galápagos rice rat to be even cuter.

The marine iguana project on Isla Santa Fé had been running for a decade before I took over. Andrew Laurie, a researcher from the University of Cambridge, had started it. Then it continued with support from the Max Planck Institute for Behavioral Physiology in Seewiesen, Germany, where I started my PhD. I can't speak to what happened before I arrived, but when we set up camp about 160 feet (50 m) up from the high tide line, I think it's safe to say it was Mardi Gras for rice rats. How they must have partied. Finally, their toys were back!

The rice rats benefited enormously from our presence. Even though we didn't purposely feed the wildlife or do anything else other than just be there, there was always the odd crumb that fell to the ground or overlooked scraps of food on the table inside our communal tent. When my colleagues were around, the rice rats were careful. They seemed to know that, at least in the early days, they would get rocks thrown at them. Not that the rocks ever hit them—the rats were way too fast—but they were still wary.

When I was alone in the tent, however, they knew I wouldn't do anything to them. They would come inside, run over me, and sit on my notebooks. I would have to poke them with my pen to get them off so I could write. They were a bit like my cat at home. The rice rats were also a constant source of amusement, and I sometimes felt I was watching a circus act. As they scurried around the table trying to find the last scraps left over from our meals, the mockingbirds arrived. The mockingbirds on Isla Santa Fé are fearsome little creatures. They weigh about the same as the rats, but they have huge curved beaks they can wield like powerful swords. If a mockingbird discovered a rice rat on the table, it would dive-bomb the rat and try to stab it with its beak. The rat would then leap a few feet into the air, maneuvering itself so it landed on the mockingbird's back and prevented the bird from scooping up the last crumb. The mockingbird would shake off the rat and fly away, leaving the rat to eat the crumb. At night, we often had short-eared owls around the camp because they knew that the rice rats liked the camp and they also knew that I liked rice rats and wouldn't chase the rats away.

Even if we tried to have as little impact on nature and animals as possible while working on an uninhabited, pristine Galápagos island, I realized that wherever humans are, there's a massive disruption of the natural ecosystem. Even though all we provided was a little shade and shelter and a few crumbs, our camp was like a raft in the ocean for our rice rats. It was a concentration point where things were happening, and those are the points where life aggregates.

After a while I knew all twenty-eight rice rats that seemed to be part of the local population. There appeared to be a lot more because they were everywhere all the time, but like any good shepherd—and a shepherd might be dealing with a couple of hundred sheep and not just twenty-eight rice rats—I knew my individuals by behavior, size, and appearance. I realized that all the rice rats had different personalities. Some were adventurous, some were cautious, and some were particularly fast.

I was usually the last person to leave the communal tent at night. I would shut off the lights and walk back to my private tent. When I first arrived on the island, I would zip up my tent because I felt a closed tent would help me get a better night's sleep. But I stopped doing this when I realized that whenever there was anything that smelled inside the tent—and there was always something—it was the signal for the rats to "go for it." It made no difference to the rats whether we had nylon tents or linen tents, they just went right through the tent material. Whenever I closed the tent, they gnawed their own entry holes. Not only that, but after chewing their small holes to get in, they would sometimes not find their holes to get out, which meant either they stayed in the tent making noises or they nibbled another hole as an exit. After a while I learned it was better to leave my tent open if I didn't want my home away from home to end up looking like a piece of Swiss cheese. I also discovered that if I left my tent open, the rats usually didn't bother to come in. I guess they reckoned if I was leaving my tent open, there was probably nothing interesting inside.

The disadvantage of keeping my tent open, however, was that about twice a season, one of the huge local centipedes would crawl in unnoticed. Like every creature in the Galápagos, they are generally tame and good natured, except if you step on them or lie down on them. If you do that, they will bite, and their bite is pretty painful. Thankfully, I never got bitten, even though on more than one occasion I did have to chase a foot-long centipede out of my tent.

One night it was so hot that I was not using my blanket and was just lying naked on my mattress. I woke up and felt an intense pain in my butt. When I grabbed my butt, I saw that my hand was covered in blood. I realized there was a rice rat in the tent. I switched on my headlamp and saw that it had blood on its face. Instead of nibbling at my toothpaste or something else that smelled delicious, such as my world radio receiver—because apparently plastic is incredibly attractive to them—this rat had just bitten me on the backside. It hadn't done this because it was mentally or physically sick; it was just being a rice rat on Santa Fé, where the little rats like to taste everything because chewing is fun and who knows what might be edible. I was shocked but I didn't hold it against the rat. It was just being a rat, after all, and the local population and I continued to be friends.

One week I happened to be the sole camp guardian when the others went back to the main island to get supplies and a break from fieldwork. I waved them goodbye, walked back to camp, and sat down at the table in the main tent, which was open on both sides and looked out over the ocean. I had been sitting there for about ten minutes, drinking a cup of tea, when—unusually for them because they had always been active mostly at night—the rice rats appeared. And it wasn't just one rat. There were two, then four, eight, ten, fifteen, twenty. At that point, I stopped counting. The rats were just hanging out in camp with me. They were running over the table. They were battling with the mockingbirds, but now they had the numerical advantage, no mockingbird stood a chance against them. They were running up my arms, sitting on my shoulders, climbing on top of my head, getting into my hair. I just could not believe it. I have no proof this happened. At the time there were no cell phones to take selfies of me sitting there with the rats all over me. All I have are my memories, but they are as true as anything.

What happened that night was even stranger. Everything was calm and eventually I went to my tent. But before I did, I looked around—and looked into the eyes of twenty short-eared owls. Two of them were sitting on top of my tent and another two were actually sitting inside it. I had heard about hermits in Bhutan who sit quietly and have a special relationship with animals, and about Francis of Assisi in the Christian culture—patron saint of animals and ecology—who supposedly talked to all the animals around him. But I couldn't believe this was happening to me.

The animals must immediately have realized that I was alone and that I wouldn't harm any of them. I wouldn't throw stones at them, I wouldn't shout at them, I wouldn't wave my arms around at them, and I wouldn't chase them. They must have observed us from afar and then judged not only on their own but also as a group that me being alone was a completely different situation compared to four people being in camp. The animals' behavior continued for the week.

It got to a point where one of the rats—the one that had bitten me on the backside—apparently developed a taste for biting me. When I was sitting at the table working on my data entries, it would come and nibble my toes. When I stretched back to relax, putting my arms behind me on the wooden boxes we used as places to sit, it would come and nibble my fingers. It's not that it drew blood every time, but at some point it wasn't fun anymore. So I caught the rat with my hand, grabbed the scruff of its neck like a parent mammal would do, and put it into a little cloth bag. I decided to carry it far away to a different rat colony.

I walked for probably twenty minutes over rocky terrain to the end of our observation area for marine iguanas, and then continued for another ten minutes just to be sure that this rice rat would never come back to camp and teach the others how to nibble on humans. Then I went down to the beach for a couple of hours to do more observations in our marine iguana colony. After that, I walked back to camp, made myself a cup of tea, and as you probably have already guessed, the little rat was back. It sat itself down on the table and began to eat crumbs right in front of me. It must have immediately turned around when I took a detour to the beach to observe the iguanas. The little rat must have been on the move all the time, jumping, running, flying back to camp to be with its buddies and its group. Curiously, the little rat stopped its biting behavior after its excursion—maybe it understood.

Now, thirty years on, the rat's behavior doesn't surprise me. There have been so many experiments using homing pigeons or other species, where animals are taken to a different place and generally know how to get back. We also know this from dogs or cats, which sometimes find their way back to their owners even hundreds of miles away. And we have experience with so-called problem animals. Consider kudus or hyenas that have been caught raiding picnic sites in South Africa's Kruger National Park and have been relocated to new areas far away. Some stay put, but others make a beeline for the spot where they were picked up and show up in that place so quickly that nobody can believe they're back.

As I watched the little rat delicately dining on crumbs in front of me, questions swirled in my head. How do animals judge people? How do they distinguish between an individual on their own and that same individual as part of a group? How do they judge how different individuals in the group act when some throw stones and others do not? And then there is the homing instinct. How do animals and birds know where they are? How do they know how to get back to a place they love or to companions they feel comfortable with? I knew we still had so much to learn about how animals process their environment. Only by tracking individuals during those interesting times in their lives would we get closer to the answers.

8 The Long March to ICARUS

IN THE TWO YEARS after watching the sunset in Gamboa, my team at Princeton had made a lot of progress thinking about a global animal observation system involving bio-telemetry tags weighing four-hundredths of an ounce (1 g) or less. In 2003, we believed we now had a viable proposal. If we could mount an external antenna on the International Space Station (ISS) with a good noise filter to screen out all the other frequencies we didn't want to receive, we could track the tags a few times every day during the flyover of the ISS.

Although biologists like us realized back then that bio-diversity loss and climate change had become global issues, it wasn't on the radar for many other people yet. Naively, I thought we could convince NASA that we had a fabulous idea.

I attended a major meeting on space science and technology in Houston, where I had the opportunity to talk to many delegates from different space organizations and different areas of expertise within NASA itself. One of the managers gave me a suggestion: "Go over to the NASA Institute of Advanced Concepts. They are the right people for an idea like this."

To me this sounded like a great idea. The Institute of Advanced Concepts. That was exactly where I thought our project belonged. I talked to the head of the institute and explained our concept and what we would need to make it happen. He thought it was all very interesting, and I felt I had finally found a home for our idea. Talking it over with him a little further, it turned out that the other major idea the Institute for Advanced Concepts was contemplating was that of a space elevator. And then it dawned on me: *Oh my God! They think tracking small songbirds, bats, and insects from space is as impossible as building an elevator in space.* I left the meeting, went back to my hotel room, shut down my computer and cell phone, put my feet up, and just sat there. Was that it?

For many of us in science, the survival instinct kicks in to rescue ideas no one else believes in. I decided my approach had been fundamentally flawed, and I tried to come up with something more strategic. First, we needed a name for the idea, preferably an acronym. Most major projects live by acronyms. It had to be an out-there acronym based on a familiar idea that everyone thought was stupid and bound to fail, an idea that had failed before and probably would fail again in the future. But the acronym should also contain the seeds of hope and great promise. Icarus! Somebody who flew too close to the sun with equipment that wasn't made for the rigors of his adventure. Although the idea was great and the outcome would have been amazing, the project failed miserably and is now known for its failure. But I stuck with it; I liked the various connotations.

I had the acronym and now it needed to be filled. We needed to make clear this was to be a large-scale movement built from the bottom up to further science. Radio astronomy had pointed the way with its VLA, a group of small telescopes joined to create one giant telescope while allowing the individual telescopes to retain their ability to function independently. So, how about I for International? That was a good start. C for Challenge. No, Cooperation sounded better. A for All of us, or Aloft. Oh no, we wanted to study animals, so Animals. R was an easy one. Research because this was what we were doing. Now came the US part. But it was to be international. Union, no. Let's just make it Using Space because we want to use space to get our data. I often get asked: "How did you come up with such a stupid name?" The short answer is "It's International Cooperation for Animal Research Using Space." But if somebody asks me over beer, I will tell the story—perhaps a foolish story—of desperation and of hope that a system that looked as though it was doomed to fail would eventually prevail and live up to its promise of flying close to the sun.

I returned to Princeton to corral some ideas, some troops, and some technical solutions. The timing was actually very good. The new administration had advised NASA to focus on landing more people on the moon and eventually on Mars instead of searching for other Earth-like planets in the universe. This was great for us because my colleague in the space engineering department at Princeton, Jeremy Kasdin, taught a class that designed instruments and systems to find other planets with Earth-like properties. He was frustrated by NASA's decision to shelve the idea, at least in the short term. And so he offered his satellite engineering design class as a Phase A study group to come up with innovative ideas for a space-based global animal tracking system. I asked my friend Kasper Thorup to join us. Kasper would later go on to test the same idea with a CubeSat—which is a miniaturized satellite intended for low Earth orbit—which he would design at the Technical University of Denmark.

We prepared a description of what we needed, met with the students, answered their questions to the best of our ability, and pulled everything together in a package that included all the ideas we had for how such a system should work. The students' task was to design the entire spacecraft, the antennas, and the communication system. They had to work on the question of whether a weak radio signal from a tag on an animal could be received by a satellite at an orbit altitude between 185 and 370 miles (300 and 600 km). After six months, the big day came and we attended the final presentations from the class. What an amazing moment. We had put young minds to work on a topic we thought was incredibly exciting, technologically challenging from an engineering standpoint, and firmly planted in the future. With the naivete of young minds, they were absolutely convinced all this was possible. They had the data to demonstrate that it was and workable solutions for all the challenges we had presented them with.

Using Jeremy's connections, we once again approached people at NASA headquarters and the students were invited to present there. This was probably a big wake-up call for NASA. The students had come up with solutions for problems it did not even know existed. The students had abandoned the idea of using the traditional analog, Sputnik-like transmissions and had designed a method of digital communication that was more like a simplified version of the system used by cell phones, which at the time had just switched from analog to digital transmissions.

When discussing this very exciting result with George and Bill, it turned out that we were actually behind the curve—yet again. The two pioneers had had meetings with NASA in the 1970s during which biologists and ecologists around the country had discussed the idea of a space-based animal tracking system. However, at the time it was assumed that the communications would have to be analog, and analog communications would not have allowed large numbers of animals to be tracked from space. The idea presented back in the 1970s would have tracked only a small number of animals across the globe and was not attractive to NASA. NASA abandoned these ideas, perhaps partly because of the political landscape of the time. There were major crises between East and West, the Cold War, and many other aspects of remote sensing to take care of first. The engineers at NASA who were involved in the 1970s were now long retired. The new generation had no clue these ideas had been discussed thirty years earlier—and neither did we. It was actually a good thing to be behind the curve. Had we known about the decision not to go for these systems, we probably would have given up right away—or at least I might have after the meeting in Houston.

We hoped we could convince the powers that be that studying the mass movement of animals, birds, and insects across the globe was important not only for understanding life on the planet, but also for preserving it and for preserving the ecosystem services all these species provide. Animals living and moving around on the planet provide trillions of dollars of these services, and if we think we can live without paying attention to and conserving these animals, we are probably doomed for failure.

But years of frustration ensued. Enthusiasm for a global animal tracking system was shared by some, but everyone was busy and there was as yet no community of people to support the idea and push it forward. There were no formal channels we could use to get our idea closer to implementation and funding. There was nobody at the National Science Foundation or National Academy of Sciences who thought the topic was interesting enough to put it on the agenda for the next decadal survey, which is a survey that identifies the big challenges and research needs for the upcoming decade. But there was still hope.

The good thing at a university like Princeton is that people are not necessarily doing better research, but they are more likely to be encouraged to pursue innovative ideas. In most cases nothing comes of these ideas, but the few that succeed make this approach worthwhile. My excellent senior colleague Steve Pacala and I shared an advisory role for grad students. We sat in on a committee meeting where a grad student said that to solve a particular problem, he would need to track a large number of animals for a long time over an extended area. Steve's response was "Well, why don't you put a receiver onto a satellite and track these animals from space?" It was a bold statement and I couldn't have agreed more.

A few days later I met with Steve again and told him this was exactly what we were planning, but it was turning out to be a little more complicated than I had hoped. Steve said he had a similar idea to build a satellite to track carbon emissions to understand their influence on global climate. Although I agreed tracking carbon was a really important question, I obviously thought that tracking animals from space was much more important. Half-jokingly, I bet with Steve that our system would be in orbit earlier than his. Needless to say, Steve's satellite was up two and a half years later and ours about fifteen years later. Nevertheless, the boldness Steve had displayed was encouraging because it showed that if you are convinced, and if you can convince others, you might succeed.

I was also learning that either you need a formal process or you need to join forces with a formal process. I thought that there must be ways to apply for funds for an experimental animal tracking system, and eventually such an opportunity did indeed come up. In 2006, there was a call from the European Space Agency as part of their European Programme for Life and Physical Sciences in Space (ELIPS) for proposals to run scientific experiments on the ISS. That looked like an ideal opportunity to me. The proposals were to be evaluated by the European Research Council. We had already run several workshops with our team of grad students and colleagues—especially Roland Kays, Kasper Thorup, and Meg Crofoot—to come up with overarching ideas for bold research projects that would only be possible using a global animal tracking system from space. Our discussions had resulted in a white paper listing about forty projects we thought would be transformative.

It is interesting to look back on these ideas and realize that some we thought were pertinent at the time remain unsolved even today; big ideas sometimes take more than one generation of scientists to be resolved. We also realized that although there were big ideas behind them, most of the projects were somewhat constrained in their scope and many were fairly simple extensions of what was already being done. We felt our project ideas were limited more by our mental scope as a research community than by what could be achieved if we had a global instrument at our disposal. We simply were not trained in thinking globally and on such a large scale; mostly we thought like traditional ecologists trained in designing experiments to observe a small number of animals over a limited area.

We were not familiar with what the European Space Agency might expect because to us it was just a name. We anticipated outright rejection of our ideas because there are often people behind these calls for proposals who have their own agendas, desires, and plans. When you submit a proposal, you just hope that some of your ideas resonate with the team inside the agency, but the chances are usually slim. A few weeks after submitting our proposal I received what I thought was a spam email. It was in the Cyrillic alphabet and had some Russian name attached to it. I was actually amazed that it hadn't automatically gotten transferred to my spam folder. Despite having some misgivings, I opened it. It was from the Russian member of the broader evaluation team for the ELIPS program, Mikhail Belyaev. He thought our idea was good but it would be better to do it in collaboration with Roscosmos, the Russian space agency, because, he wrote, they are "the best and quickest" in implementing space solutions. He was right. For the longest time the Soviet Union had been the only nation that could launch astronauts into orbit. I thought this was a very exciting proposal and wrote to Belyaev that I would be happy to discuss it.

A few months later, we received a formal evaluation from the European Research Council: They thought the scientific ideas were important and timely, the international team was excellent, and the structure of both the team and the plans was good. However, the engineers on the review panel suggested that it would very likely be impossible to actually implement such a system. In short: A great idea, a great team, a great implementation plan, but most likely technically impossible. Full support, but no funding. This was a major blow, of course, but also a firm acknowledgment that we were onto something big.

Schloss Ringberg

9 Switching Back to Europe

A COMBINATION OF PERSONAL and professional factors, particularly the positive feedback on our ICARUS ideas, made me consider a move from the New World academic system back to the Old World system. NASA was still not really interested in global animal tracking, and the positive review of our proposal by the European Research Council and the keen interest on the Russian side made it more likely that Europe would be the best place to implement our ideas. The possibility of long-term funding from the Max Planck Society was also a big draw, and in 2008, I accepted a position as director of the Max Planck Institute for Ornithology based in Radolfzell, Germany, and another as honorary professor of ornithology at the nearby University of Konstanz.

After I moved back to Germany, it was a completely new ball game. For the first months and years, I was largely consumed by setting up new research connections and trying to make inroads into the decision-making systems and agencies that needed to be convinced that ICARUS was important. What followed was half a decade of meeting after meeting after meeting and many, many dead ends. I have a number of guiding mottos. The first is that if you don't try, you won't succeed. This sounds simple, but most projects are abandoned and fail at the outset because the task seems too vast to ever be achieved. The second is that you have to be lucky, but when you are, you have to follow up. And the third is that you need to work to find the right allies, but you never know where you will find your true allies.

Right away, we set about broadening our scientific base by organizing a meeting about the future of talking to animals on a global scale. We met in the mountains of Bavaria in one of those fantastical Disney-looking castles built by eccentrics who wanted to create something beautiful in a wild landscape. Situated above the Tegernsee (a lake), Ringberg looks like a smaller version of Neuschwanstein, the famous castle built by the Bavarian king Ludwig II. Ringberg enveloped attendees in a slightly fairy-tale atmosphere that allowed everyone to discuss things that normally they wouldn't dare.

We invited an international team that combined some of the most experienced biologists we knew along with younger minds. Our goal was to create a road map to lead us to where we wanted to be with such a system in the future. Those we initially considered the old farts, the established traditionalists in science, turned out to be thirty years ahead of us. They had challenged accepted wisdom in their generation and now helped us hand over the challenge to the next generation. There were colleagues from the National Geographic Society; staff from the U.S. Fish and Wildlife Service; researchers working in the marine, terrestrial, and aerial realms; mechanical, communication, and space engineers; and veterinarians. All had been hand selected for their big-picture thinking, their visionary approaches, and their interest in advancing a wide-ranging system and not just their own personal agendas. It was a stimulating meeting that kept us going for a good decade. Whenever frustrations took over, we thought about Ringberg and the collegial and entrepreneurial spirit that had prevailed.

During the decade after my return to Europe, we had some incredible luck. The most pivotal event was a May 2009 meeting of the Bavarian Academy of Sciences and Humanities. The topic was navigation. Many presentations were about obvious topics, such as global navigation using GPS comparing Russia's GLONASS satellite navigation system with the start of the European Galileo system. This was followed by discussions of GPS in self-driving cars, navigation systems for pedestrians in cities using cell phones, navigation systems using motion sensors and dead reckoning, and other technologies.

At that time many attendees still considered some of these approaches as ridiculously unnecessary. My task was to push the envelope even further with a talk about navigation systems and animals. As I did in all the public seminars I gave at the time, I highlighted how important it was to have little GPS devices on animals to tell us about what decisions they make in the wild, what knowledge animals accumulate over their lifetimes, and all the places they have been. If we didn't know that, we would never know why they make certain decisions such as flying around rather than over a mountain, swimming in one ocean but not another, or migrating toward what appears to them to be greener grass far from their homes.

I could tell from the reaction that the minds of the participants had not made it past the highly controversial discussion on pedestrian navigation with cell phones. Most people wore slightly glazed expressions and seemed more interested in the buffet that beckoned after my talk. I left right after my talk and did what you do in Bavaria. I looked for a beer garden where I could relax and lower my level of frustration with a beer or two. However, it was a drizzly day, nobody was out, and after an hour and a half or so of searching, I went back to the academy to salvage some scraps from what had apparently been a wonderful dinner.

By that time, everybody had left except for a couple of the old farts, who were standing around a little table. One of them, who by the look of him had retired a long time ago, waved me over. "Aren't you the one who talked about animal navigation? That was new and interesting." I thought he only wanted to be nice and chatty, but it turned out that he was one of Germany's most accomplished space engineers, Phil Hartl. He had done his postdoc at NASA's Jet Propulsion Laboratory in California. In his cubicle work space at the time, he had sat back-to-back with Andrew Viterbi, the guy who devised the Viterbi algorithm, which is almost the Bible of space communication systems. It's a bit complicated to explain here, but trust me, it's a big deal.

Phil and I just happened to hit it off perfectly. He was mentally as young and enthusiastic as George and Bill, but his expertise was somewhat different from theirs. Phil had been one of the masterminds behind radar observations from space, but since he was a geographer by training, he had also figured out how to use gravity missions from space to solve one problem everyone had thought was unsolvable: how to measure changes in the global distribution of groundwater. Phil had solved this issue by getting one satellite to fly about 60 miles (100 km) behind another. Then he measured distortions in their orbits, which are caused by gravitational changes on the ground. If the two satellites flew over the same place and their gravitational forces were different, either an incredibly large crowd of people had suddenly gathered at that spot or something was going on with the amount of water present. As we can measure differences in surface water, anything unaccounted for in these calculations must be due to differences in groundwater. I guess nothing more needs to be said about how important these measurements are to what is happening on our planet during the period of climate change we are experiencing.

After we chatted, Phil invited me to present my ideas in a five-minute talk at a beer hour in Munich. Now, this was not your regular beer hour. This one truly earned its name: the beer hour of advanced space concepts. There were former heads of the European Space Agency who had been responsible for making the iss happen at a time when nobody believed it would ever come to pass, and people who had started the European Ariane rocket program. There were people working on the European Galileo system. Basically, all the people who attended either had worked on or were working on significant aspects of the European space system. Being invited to speak at this beer hour scared me more than any job interview I had ever done or any seminar I had ever given. All I had was the power of my words for five minutes. But if you have the right audience, you can give the right talk.

I explained to a bunch of the highest-level engineers in Europe that humans had lost their connection to nature during the time of the industrial revolution. This was a fatal mistake for humankind because we share our planet with plants and animals and can survive only if all of us are doing well. Plants are covered by satellite remote sensing systems, but animals are not. We need to understand animals to bring people back to nature, and we need to give animals a voice because only if we learn from them can we survive with them in the future. This minimal investment in novel technology would create a huge payoff for humans in terms of knowledge, in terms of beauty, and, last but not least, in terms of the conservation of beloved and endangered animals.

The term Internet of Things, which is commonplace today, had not yet been coined. The Internet of Things refers to computing devices embedded into everyday objects so they can exchange data with other devices and systems. Components of the Internet of Things include smart cars, smart fridges, and smartphones. What I was envisioning back then was an Internet of Animals that would lead to a similar web of integrated information exchange in the natural world. In my presentation, I described a global network of information flowing from multiple living nodes that would allow us to share the wisdom of the most intelligent sensors that will ever exist on this planet and, as far as we know, in the universe: animals.

Animals are so diverse that they all have different feelings, senses, and social skills, both as species and as individuals.

The sum of their knowledge is superior to the sum of any knowledge we will ever gather from human-made systems. It is proven knowledge, sustainable knowledge that has prevailed through the eons, through times of incredible global and climate change, and through dramas of unimaginable extent—meteorite showers, global winters after volcanic eruptions, and outbreaks of deadly diseases.

The applause was short, quiet, and followed by a few questions. Then everybody went back to discussing the future of the space station, the future of rockets, and the human dynamics integral to international space relationships. But there was one person who came up to me and asked me to write my ideas down and send them to him. I didn't know who he was at the time, but it turned out that he was Christoph Hohage, director of space projects at DLR, the German (not European) space agency.

Hohage is a straight-talking guy who articulates his opinions clearly and fights for his convictions. He is one of those people you want to have on your side. He is the kind of person who will tell you when you are woefully off track and support you when you are on target. He is also the kind of person who is incredibly critical, both of himself and of others. We wrote a proposal for a small pilot study to outline the assumptions, constraints, and framework for an experimental global animal tracking system using space assets. It allowed Phil Hartl to work with our team, particularly with Uschi Müller, the all-important coordinator of the ICARUS project, who started working with us in 2009. Hohage was also the person who recommended we talk to one of the new space start-up companies in Germany, because our project was still too small and insignificant for the big players in the space industry. Bernhard Doll, the head of SpaceTech, the German space start-up company we partnered with, was actually one of the most accomplished space engineers in Germany, having contributed to all major successful space missions in his previous function as head engineer of one of the big space companies.

Another lucky break was that the head of administration at the Max Planck Institute for Ornithology, Thomas Dzionsko, had previously worked at the Max Planck Institute for Extraterrestrial Physics. One of my scientific colleagues there, Gregor Morfill, had just developed applications for cold plasma and he had carried out his initial research in conditions of microgravity. The important part here is that Gregor had exceedingly good connections with Russian colleagues and that country has a tradition of training extremely good physicists. Gregor had worked for a long time with a Russian professor called Vladimir Fortov, who at the time had been the head of the Russian Academy of Sciences. Their relationship showed how well joint projects work across political divides—if people want to collaborate and if they are willing to tolerate each other's peculiarities. With Gregor's encouragement, we revived our contacts with Russia and set up a joint engineering team. While our team was extremely strong on electronics and novel communication technology, the Russian team had a proven track record in making complex systems work in space. Tradition and innovation, chance and necessity, naivete and experience. When those come together in a team, magic happens.

We also had the good fortune to be working with a team of young communication engineers who had just formed their own start-up company, INRADIOS, and were happy to take on the challenge of devising a digital system that could track animals from space and create a web of information from the data. The people responsible for reviewing the implementation of our project had sincere doubts that a digital system would work because digital signals are so much weaker and harder to process. They urged us to stick with the more traditional route of analog communication. But our communication engineers were naive and confident and thought they could make it work—which, to their great credit, they did.

The white
stork Hansi

10 Who's in Charge?

AS WE WORKED THROUGH the practical issues involved in getting ICARUS off the ground, I thought more about how personal the communication between humans and animals can be and how incredibly interesting the stories would be if animals could be the ones telling them instead of us.

In literature and popular writings, it is always clear that humans domesticated animals. There are specific times and places where this supposedly happened, be it in Mesopotamia, where people tamed animals for meat, milk, and hides; in ancient Egypt, where dogs and cats, baboons, fish, and gazelles shared homes as pets; or fifteen thousand years ago when nomadic hunter-gatherers first domesticated the gray wolf—perhaps. But animal tracking sometimes tells a story that should make us rethink our hubris that humans are the ones in charge and that animals are just the recipients of our domestication efforts. If we listened to how animals told their stories, we might learn things that would change our perspective.

George Swenson, the former radio astronomer, always said that communication with extraterrestrial life-forms would be the most important achievement of humankind. In his thinking, we would get insights into other places in the universe and, most importantly, an outside view of ourselves. It would be like when you first hold a mirror in front of your face and you realize how other people see you. But before relying on extraterrestrials, I thought, why not make use of what we already have right here on our own planet? Why not take advantage of the outside view of ourselves that animals could give us if we allowed them to communicate with us? Why not develop a system that would give animals even more of a chance to give us their side of the story?

Take the story of the white stork Hansi. As background, I should say that before we began our stork project, I was told by most people that we should not continue to study white storks because we wouldn't learn much. They were the first birds to be banded in Germany and had already been studied for about 115 years. They were also the first birds to be the subject of citizen science. People were encouraged to report their sightings of banded storks using postcards mailed to a central, bird-banding observatory, which was an amazing initiative at the time (circa 1905). People thought everything was known about white storks. Or, to put it another way, it was suggested that I focus my scientific investigations on a subject that would provide completely new answers rather than on getting a slightly more detailed understanding of what we already knew.

Well, for explorers and scientists, statements like that have always been an interesting challenge. Whenever somebody says "everything is known," they are clearly missing the bigger picture. Tradition begets true innovation: first you have to know your skills before you can invent new ones. And so, my team at the Max Planck Institute for Ornithology continued to work on storks, and in the summer of 2013 we fitted tags to a whole bunch of young storks before they migrated from their breeding grounds in eastern Germany. The tags would relay information to us via the cell phone network, and tiny solar panels on the tags would keep the batteries charged.

We were working close to Loburg, a small town about 60 miles (100 km) southwest of Berlin. Those were the early days of the biologging revolution, and we thought that tracking forty young birds in one year would be an absolutely amazing achievement. But imagine asking forty people if you wanted to predict something in a human population, such as the outcome of an election. If you interviewed forty young storks out of fifty thousand, all of whom hatched that year in the same rural area of Europe, and asked them all, "What are you doing?" your ability to make predictions would be extremely limited. At the time, however, we were happy that we were able to track even forty of these young birds.

And indeed, this one study gave us amazing insights into migratory divides, where some storks fly west around the Mediterranean on their way to Africa and others fly east. We were particularly interested in how the migration patterns changed in response to global changes, including climate change. Most storks from this area in eastern Germany migrate through Czechia, Slovakia, Romania, and Bulgaria into Türkiye, before continuing through Syria, Lebanon, Israel, and Egypt into Africa to settle in wintering sites either in the Lake Chad area—Sudan, Eritrea, Ethiopia, Kenya, Uganda, Malawi, Mozambique—or all the way down in South Africa. Historically, many of the storks had migrated all the way to South Africa; recently, however, only a few of them had been flying that far. There had been a shift in water availability throughout Africa over the prior twenty years as a result of climate change. Perhaps this was why we were seeing changes in the migration patterns of birds from breeding areas in Europe. Although most of the storks chose the eastern migration route around the Mediterranean, a few of the Loburg storks flew west through the Netherlands, France, and Spain, crossing the Strait of Gibraltar to Morocco. And some of our storks used both routes into Africa, one year flying west via Gibraltar and the next year flying east via Israel.

What this tells us is that there is no genetic predisposition in storks to fly east or west. It's a social migration system. They follow everybody else. When the storks are ready to migrate in midsummer, they look around to see what the others are doing and there are always some more experienced birds that know where to go. Thus the migration patterns of our young Loburg storks are culturally determined, and young storks follow more experienced storks that lead the way.

One of this cohort of forty birds we were tracking was our stork Hansi, who was initially known only as HL430. Hansi was a particularly interesting case because he left the breeding grounds extremely late. He managed to find a few other stragglers to migrate with and on the first leg of his migration he took an excellent route toward the southeast. He flew through eastern Germany and into Czechia, where he rested for a few days. We assume that when all the other storks left to continue their journey, instead of following them Hansi decided to make a ninety-degree turn to the right, flying directly southwest on a route no other stork (that we knew of) had ever taken.

I couldn't imagine that our Hansi had flown with other storks on this southwestern leg. But I also couldn't imagine that he had flown completely on his own, because his track was too straight and too closely aligned with tracks that another species of large white bird was taking in this area: the great egret. This potential case of mistaken identity immediately caught our attention, and we wanted to find out what was going on.

As soon as Hansi stayed in one place for a day or two, which occurred southeast of Munich close to and slightly north of the Alps, but still in the foothills, I hopped on my motorcycle and rode out there to find him. Lo and behold, there he was, perched on top of a huge tower in an electric transmission line. Several egrets were on the ground in the field around the tower. Hansi had apparently encountered some local storks too, as his location was only a couple of miles from the next known group of his conspecifics, and at least one of those storks was also in this field. Hansi spent a few days in this beautiful low-elevation meadow. We expected him to join up with the local storks and fly east along the Alps, following a route along the curve of the mountain range as it turns south. But this was not what Hansi decided to do.

Apparently, Hansi was convinced that "his" egrets were better guides. He flew with them straight south—no self-respecting stork would do this on his own—into a valley, crossed a minor mountain pass, and then continued on almost to the skiing town of Garmisch. From there, he took another bizarre turn, flying over another pass into the valley where Innsbruck is located. Then he flew directly toward the Ötztal. This is the valley where Ötzi the Iceman would have ended up five thousand years ago as he made his escape from Italy north into Austria had he not been killed when crossing the mountains from the south.

For us, this was an amazing discovery. Instead of joining the other storks to fly to Türkiye and Israel and then farther south into the warmth of Africa, Hansi was apparently flying with the egrets directly into the Alps. We now wanted to get a good look at where Hansi was hanging out. Like all of us, my coworker Heidi Schmid was desperate to find Hansi, and so she drove four hours from our institute to his hangout location. She found him standing in a mountain valley meadow with several great egrets. He was happily foraging and apparently in good shape. Now all bets were off as to what his next move could be.

But then the weather turned bad and Hansi fell silent. As is typical for the Alpine region in late fall, it can be almost completely overcast for weeks and the days are really short. The solar panels on Hansi's tracking tag were no longer generating the energy the tag needed to send us data via the cell phone network. Even though it was not transmitting, the tag would still collect enough data to record a few locations per day as well as Hansi's body acceleration, which would allow us to decode his behavior after the fact if we could retrieve the tag to download the data or if the tag ever got enough sunlight again to recharge the battery and resume transmissions. In the meantime, we had no idea what had happened to our beloved (and apparently weird and wonderful) stork. For all we knew, Hansi could be dead. Or the tracking tag could be damaged. Or maybe the bad weather wouldn't last long and we would eventually get another message from him. But that seemed highly unlikely.

When Hansi was silent for two months we almost decided to shut down his cell phone contract, because individual cell phones for forty kids—our forty storks—are pretty expensive, especially when you are paying international roaming charges for calls from the Middle East and Africa. But thankfully, we were not efficient enough to cancel the contract immediately. We couldn't believe our luck when we received a data point on December 10 from northeastern Bavaria. The data came from Altötting, a village known for its religious pilgrimages. I honestly thought that somebody had found Hansi's tag and brought it to the small community during a pilgrimage. This was the only reason I could think of that would explain why we would get a message from Hansi's tag from the middle of a village in this area in December. Our stork should have been far away to the south by now, along with all of the rest of his colleagues.

As soon as I got that single data point, I called Heidi Schmid again. "I know this sounds ridiculous, but can you drive to Altötting, please? Not for a pilgrimage, which you may do if you want to, but to find out whether a villager or a farmer has found Hansi's tracking tag. Perhaps they are keeping it on a windowsill somewhere in the middle of the village. Please try to leave immediately, because right now it's sunny and we don't know whether we will get any more messages from the tag."

So, Heidi took another long drive (five hours this time) and arrived in the village right at the time when the kids were coming home from school. She saw a young boy walking along the road and asked him, "Have you by any chance seen a large white bird here?" And the little boy said, "Sure. He's in the field behind the farmhouse just over there. He adopted the farmer's family." Heidi drove over and saw our stork Hansi practically in the farmyard. She knocked at the farmhouse. "Is this stork staying here?" she asked. "Are you seeing it a lot? Is it visiting you?" And the grandma, the old farmer, said, "Oh, yes. We called the stork Hansi because it arrived just when my daughter became pregnant a month ago. We plan to call our new grandson Hansi, so we thought we might as well call the stork by the same name."

Heidi took some pictures and asked the family whether we could come back with a small team to make further observations. A few days later, our entire stork team drove there, and we couldn't believe what we saw. The young boy had been right. The stork had adopted the farmer's family. It was a farm that had transitioned to the organic farming of sheep some years ago. The beautiful old farmhouse with its lovely yard was in the middle of the village. There was an orchard full of apple trees, some henhouses, and all the other things you might expect on a farm. When we walked around the corner of the building toward the garden, Hansi was standing in a shallow basin filled with warm water.

Grandma told us that she really appreciated that Hansi liked warm water for his feet, because that was her favorite daily treat for herself, as well. Every day, Grandma brought out a warm water bath for Hansi, so he could stand in the garden and have warm feet. She had also learned that Hansi preferred his sheep liver minced. Grandma was not only providing a water bath but also a dish of finely cut sheep liver every day. She served other goodies for Hansi every day throughout the winter too. Grandma actually thought she had to do all this because the stork had brought her another grandson. After observing Hansi for a while as he warmed his feet, we got cold ourselves and were invited by the family to join them inside for a traditional Bavarian dish of white sausages (made from organically farmed sheep).

We sat down in a dining room with a large window and a glass door that led out to the terrace. From there, we could observe Hansi from inside. What happened next was so unbelievable I am glad my colleagues were there with me to witness it. Hansi stepped out of the water bath—perhaps it had become a little too cool for him—and walked up to the glass door. He extended his long red beak and knocked several times on the glass. Grandma stood up, explaining to us—and apologizing to Hansi—that she was late. She got a dish of minced sheep liver and put it out on the terrace for him. He did not move away from her. Hansi apparently knew what was to come: excellent organically grown food for a stork in midwinter. Grandma said a few nice words to Hansi, closed the door, and continued to chat with us. Except that we were speechless. We had just witnessed, with our own eyes and ears, that the young white stork HL430, aka Hansi, had actively adopted a farmer's family and had asked for food from his human friends because Grandma's meal delivery service was a little late owing to our arrival.

This was perhaps one of the most transformative experiences of my early tracking career. Hansi making that ninety-degree wrong turn, following egrets instead of storks (presumably), and then adopting a human family was something that just blew my mind. Tracking animals with wearables for wildlife and thus giving them a way of telling their own stories had turned my world upside down.

The rest of Hansi's story was both beautiful and sad. Hansi spent the entire winter with the family, but then eventually took a trip to the northwest in the middle of March—when his urge to migrate north apparently kicked in—only to make a 180-degree turn at the end of March to fly back toward the southeast, where he probably met up with other storks returning from Africa flying northwest, headed straight for Hansi's birthplace in Loburg in eastern Germany. However, when he hit the spot where he had taken the southwest turn with the egrets the previous fall, he once again made a 90-degree turn and headed southwest, taking almost the same route he had taken the previous year. This time he ended up near Salzburg in mid-April. He stayed there for a few days, but then searched along the northern part of the Alps, presumably trying to find his old favorite location in the Ötztal, which he couldn't.

Hansi reversed direction and spent the rest of the spring in Salzburg, before flying south into another mountain valley in June. On June 20, Hansi's urge to migrate apparently surfaced again and he went on a southeasterly voyage, following exactly the same route he had taken the previous year, except this time he must have joined his old stork buddies from Loburg on their journey toward Istanbul and the Black Sea. He flew with them through Türkiye into Lebanon,

Israel, and Egypt, and made it all the way to Lake Chad. From there he went east toward the Horn of Africa, but then eventually went missing in the disputed area between Ethiopia and Eritrea. He might have been caught by a predator and eaten, or he might have hit an electric transmission line and been electrocuted. More likely, though, he was hunted. But I don't know for sure.

With the tags we were using at the time, we had no idea what happened after our tagged animals vanished outside cell phone range. This is why we needed our ICARUS global observation system, our Internet of Things system for wildlife, so we could keep track of the wild animals that domesticate us from time to time. In this particular case, HL430, aka Hansi, was one of the vanguard—like all the others in this cohort—and he told us a very, very special story about how one bird dealt with its life challenges. Animals are certainly a lot more inventive and adventurous than we could ever have imagined.

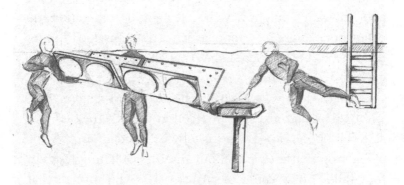

11

The ICARUS Design Starts

THANKFULLY, WE HAD no clue how many pitfalls there would be on the road to a satellite system. How many moments of despair, how many times when we desperately wanted to give up, because the whole process had become so exquisitely frustrating that we just couldn't stand it anymore. And this was not only for a month or two but for an entire decade. What outweighed the negative aspects were the moments of absolute joy: when we got the rocket launched, the receiver into orbit, the spacewalk going. It was unbelievable how rewarding these moments were when we were achieving things that had never been achieved in this field before.

But I am getting ahead of myself. There had been that naive phase of thinking *We can get this done in two years.* It was now 2012. We finally had some funding from the

German space agency, we had a project team, and we could start. At first it was hard to get the engineers and the biologists to understand one another. Each field had its own language and its own way of moving forward. We found the only way to achieve any real progress was to get the groups to meet in person. That and the fact that all of us were equally keen to get the system up and running.

After about a year, we had our first design review by our funding organization. At the time we didn't realize the future of the project rested on the outcome. Fortunately, we had one amazing reviewer, Andreas Knopp, who was as willing as we were to embrace innovation while still having a strong grounding in traditional technology—a perfect combination for us. He agreed with our communication engineering team that although it sounded marginal or perhaps impossible to establish coded communication between a tiny tag and a space asset—in our case, a receiving antenna on the ISS—we should still go for it. This was exciting because now we could have our first meeting in Moscow.

When we entered the buildings where the engineering design had been done to send the first humans into space, we felt as though we were making a pilgrimage to a holy site or having an audience at the Vatican. It was all a little bit like in the old James Bond movies with high security and the typical KGB types hanging around, who were introduced as special observers but who everybody knew were security agents. Old chairs and furniture, and the smell of communism still hanging in the curtains.

We sat around a huge conference table and noticed that everybody on the Russian side was a little nervous. The interpreters who were always there to translate, even though our colleagues understood English very well, put on their smiley faces and chatted a bit with us. It was actually really nice to have interpreters, because when they are there, you have to formulate short statements and short answers or they can't really translate. And you have time in between the statements to think about what to say next. It also means that a meeting that might otherwise take three hours is over in thirty minutes because you don't chat, go off on tangents, or do a lot of the usual blah-blah-blah that never adds anything of substance. Then the door opened and an old man entered. Our Russian colleagues stood up and naturally we stood up too, because we had no idea what was going on.

We then learned why our Russian colleagues had been nervous. As it turned out, this man was the space engineer Viktor Pavlovich Legostaev, the general designer of RSC Energia (a subsidiary of the United Rocket and Space Corporation) and the main architect of the Russian human spaceflight system. As a young engineer, he had constructed the re-entry system for Uri Gagarin, the first human in space. The guy looked as though his mind was firmly fixed on what might be about to appear over the horizon. The only other time I have seen this look is on marine captains standing on the bridge of their vessels scanning the open ocean and anticipating things to come. This way of looking at the world is ingrained in vessel captains, and it appeared to be ingrained in this guy, as well. He was constantly looking beyond the horizon because he wanted to know what the future would bring.

Despite his vaunted position, Legostaev was totally unassuming. We chatted through the interpreters, and I told him about our plans to set up a global system to communicate with animals using space assets. I had no clue how he would react, but I knew that the future of our project depended on his assessment of our project and, perhaps even more importantly, on his assessment of me. He was one of those people who, if he believed in a person, would allow his team to make things happen. After some minutes, he said, "I grew up on a farm and I have a dacha now. I know animals can tell you things and it will be good to understand what they say. This is especially important for Russia because we are such a huge country. This is a good project and it should have priority." He looked at me and our team for a while. It almost seemed as though he was remembering the early days of his own team when he was in a similar situation, designing a re-entry capsule for the first person in space. Then he returned his gaze to the horizon and left. Everybody stood up again. That was it.

It's not the best thing if one person decides, but it can be so much more efficient than endless committee meetings. At least the decision had been made and now the project could move forward. Legostaev's support meant that we were immune from the tumultuous world of international politics, at least for the time being. His imprimatur carried us through a decade of interactions with the Russians. We encountered lots of problems. There was the crisis in the Crimea. There were spies and activists being killed by Russian agents in the U.K. and others being incarcerated in Russia. Through all this, the advice was always the same: collaborations in space, particularly revolving around the ISS, were exempt from any issues. "Just continue what you're doing," I was told. "There'll be a time after this is all over when we'll need to bring down the tension again and a collaborative project where everyone benefits, such as ICARUS, could provide a good starting point." We also had a good politician in German chancellor Angela Merkel. It was clear that support for science and technology percolated through our ministries.

I don't remember how many meetings we had in Russia. But there were many. The Russian engineering delegation also visited us in Germany many times as we tried to design a workable system. The most challenging issues through all those years were people who had their own agendas. They claimed they were good team players working for the common good and trying to advance science and technology, but in reality they were acting out of self-interest, hoping to benefit personally in the long term from all the innovative thinking in the room. This was frustrating and discouraging because their personal agendas almost killed the project. The most difficult part of project management is when you have to kick these people out and retrench for a while to ensure the project's survival.

Russia, Russia, Russia... The meetings in Russia were good for a surprise, for a lot of fun, but also—always—for a lot of drama. The fun was to have a little bit of time to hang out in Moscow, which is an incredible city with incredible people, and perhaps at the time one of the most fashionable cities in the world, but also a city with many problems even during what were supposedly good times. It was a special treat to visit the space museum managed by RSC Energia, the human spaceflight company in Russia. It was unclear to us why all these priceless space assets were not on public display. Perhaps the Russians were still wary of spies, or perhaps they did not have the funds—or maybe the foresight—to display these treasures in a public museum. The collection included the re-entry capsule of the first man in space. We also saw the first space sauna—obviously designed by Russians—the adjustable seats in the space capsules, the engineering model for Sputnik, and many other jewels left over from the space race. And the drama? Well, it was embedded in everything.

As the process unfolded, we had to solve not only technical issues, but also a lot of administrative ones, such as who ultimately owned our receiver and antenna. When we transferred it to Russia and it went up into space, would it still be German property? If it was, did it belong to the Max Planck Society, to the German government, or to me? This was important because someday, when the space station needed to be decommissioned, there would be a huge piece of metal in orbit that might fall down onto a populated area. Even though this was exceedingly unlikely, what were the ramifications of our piece of metal falling onto, say, the Chinese embassy in Washington? Did anybody know? Might it be better to just shelve the discussion for now?

At the same time as our engineers were designing the space assets—the receiving antenna and the computer we would send up to the ISS—we also needed to start work on the ground systems: the little tags that were going to communicate information from animals into space. This was the single most important part of the project and, as it turned out, also the most difficult one to get underway. The German space agency, by default, could fund only space activities. But the tags were on the ground. The department responsible for remote sensing was focused on the physical properties of the globe, which included plant growth but not animal movements. The tags were also not equipment; as wearables for wildlife, they were consumables. This hairsplitting was a huge problem for our project. We were working with the ISS, but what we were doing was remote sensing with animals. There was no obvious funding slot for this revolutionary way of scanning the planet.

It was now critical to find funding for tag development. And again we realized how important individual people are in positions of power. This can be good, but it also can be bad. We were allowed to send a proposal to the one unit of the German space agency that could give small grants for these activities. But the person who was evaluating these proposals was a colleague of a person from another program who wanted to build these tags as part of a more wide-ranging platform that included tracking large equipment, trucks, containers, simply everything. While eventually this might turn out to be a good thing and perhaps after ten years of development this is where everything would end up, you cannot start from the most complicated and complex system and expect to get anything going in two years. This meant that our proposal was rejected. We were basically out of funds for tag development because one guy wanted to be the person who became famous for developing these tags.

All we could do was continue to rely on the long-term support of the Max Planck Society. Without it, the project would be over. It wasn't a huge source of funding—barely sufficient to get started—but it was sufficient to keep our spirits up and things going. The pushback and uphill battles seemed to be a constant in the project anyway. It was consistently underfunded for what it wanted to achieve. Space systems were usually in the high double or triple hundred million dollars and our system was barely one-tenth of that.

Another important aspect was that with bare-bones funding from a scientific institute, there was no way we could manufacture a large number of tags to supply a lot of collaborative projects around the world, which was our goal, so we decided to start our own company. This might sound ideal, but in practice it was a complete disaster because the person we installed to lead this company turned out to be one of those individuals pursuing a personal agenda. After a short while, the company went bankrupt and we had to abandon it.

We also encountered fundamental differences in the way the Russian engineers and the German engineers wanted to design the space antenna. The Russian approach, which is usually an excellent approach, was to stick with technology that had proven itself in the past. But for a new communication system like ICARUS, the first ground-to-space Internet of Things (or in our case, animals), we needed something new. We wanted to design a smaller version of the Very Large Array, the idea George had to combine single telescopes to create a composite eye through which to view the universe. The head of the German engineering team, Bernhard Doll, had two goals. The first was to build an antenna array that could be reprogrammed if the demands on it increased. The second was to build an antenna that could function as the base of its own satellite sometime in the future. At the time, no one else had thought about these design features. In the long term, they would turn out to be crucial when we needed to completely reconfigure the ICARUS communication system. But more on that later.

A problem with underfunding was that we could build only one antenna. Under normal circumstances, we would have had two systems built, one to keep on the ground so any problems occurring in space could be figured out in the laboratory, and the other one to fly to space. We opted instead for an engineering flight model, a euphemism to hide the fact that we had funds for only one antenna, and we built a neutral buoyancy model of the antenna because the cosmonauts needed to train with something to understand and practice exactly how to maneuver this huge piece of equipment during a spacewalk outside the ISS. We did build two systems for the onboard computer so we had a backup on hand, and this turned out to be incredibly lucky later on.

If it hadn't been for the Berlin airport (where construction had by that time dragged on for years), I would have judged the Russians more severely for the time it was taking them to fix a leak in the water tank they used for cosmonaut training. Cosmonauts need to be zero-buoyant on Earth to practice and test spacewalks. The only way to do this is in water wearing clunky diving suits that resemble space suits, but the water tank in Russia had been dry for the past two years because of the leak. That meant we couldn't get the antenna's neutral buoyancy model tested, and thus there was no way to prepare for a spacewalk and the continuation of the project. We tried everything. We went to the European Space Agency facility in Cologne. We contacted NASA's training facility in Houston, Texas. The German space agency got in on the act. In the end, the best place to get started turned out to be the city swimming pool in our little town of Konstanz in southern Germany. We convinced the cosmonaut trainers to come to Konstanz and check out the neutral buoyancy model in our municipal pool.

It was quite an event in town, because everyone was fairly sure there would never be another occasion on which a space antenna would be tested in the local pool. The testing went off without a hitch. We also made it a showcase for the young people we were using as our ICARUS advisers at the institute. We already had enough old people telling us what to do. What we wanted were young people with their open minds to tell us where the future lay. They were the ones who jumped into the water with the antenna and helped make the system float.

In 2012, I still thought we could be in space by 2014. But our experienced space engineers told me that was ridiculous. They said the earliest they could be ready was 2016. This was a gut punch for me because we didn't know where we would get the funding to keep the engineers employed for another two years or how to tell the scientific community that we needed more time. If I had known in 2012 how many years we still had to go after 2016, I think I would have given up.

The year 2016 came and went. We still had massive problems in many areas, but our general system architecture was working. Our antenna, conceived as a miniature Very Large Array, was working. Our digital communication system was working. "All" that was left was to get the antenna in orbit and connected to our tags.

One major challenge was to get equipment from Germany to Russia for testing and, eventually, for the launch. Most people, including me, had no idea how rigorous the quality checks are for all space systems. All functions of the computer, for example, needed to be tested by the German and Russian team in Germany, then sent to Russia to be unpacked by the German team and checked again by both the German and the Russian teams.

To make sure the computer would plug in correctly on the space station, there was a model of the ISS housed in a huge assembly hall in Moscow. Our engineers had to make sure that every cable fitted into the allocated rack in the ISS, and that the power supply had the right polarity. Simple things, any of which could go horribly wrong. But all this testing allowed us to visit a facility where very few people from the West had been before: the assembly hall where all the Soyuz rockets were lined up. Well, actually these were all remodeled tips of intercontinental ballistic missiles that were now being used to supply the space station. And there it was, finally, the model of the ISS. A backup testing system lying on the ground attached to all the cables and everything that had ever been, or soon would be, put into orbit. Except for our antenna. We couldn't leave it attached to the ground model because we had only one.

The part that amazed me most, though, was an old Russian woman sitting at an ancient wooden desk that looked as though it had been used for centuries. She had a paper notebook in front of her that was perhaps 6 inches (15 cm) thick, full of small lines and script, most of them in Cyrillic. She was the person writing down every connector, every cable, and every activity on this model of the ISS. Her notebook was an analog memory of everything that had ever been done to the ISS and installed up there on the Russian segment. If anyone ever wanted to go back and see which screw had been put into which cupboard of the ISS, and had later been moved and put somewhere else, they could go to this woman with her notebook and get a complete rundown of the ISS. She was like an analog CCTV camera, but much better. Because everything was already decoded, written down, and fixed on paper, no one had to watch the CCTV footage or rely on clever artificial intelligence to understand what had happened.

While the engineers were standing below the model of the ISS where our antenna would need to be attached, discussing the best place for it and measuring how many lengths of cable would be needed to fix the antenna to the outside of the ISS and where the cable should run, we stood on an elevated walkway with Mikhail Belyaev and watched the whole procedure. Out of the blue, Mikhail told me that his grandfather had been shot by the Bolsheviks during the 1917 October Revolution. His father had to watch this as a young boy. He also mentioned as an aside that he was always a little bit afraid that this might happen again, especially because despite all the changes in international politics, the Germans might once again be involved. He mentioned that he thought Angela Merkel, our then chancellor, was one of those colonialists who might like to reinvade Russia.

It suddenly hit me—and I should have never assumed otherwise—that there were still these generational resentments. There was no personal animosity, but there was definitely baggage piled up in the background that was coloring all our dealings and would likely continue to color them for at least another couple of generations. This was good to know, because it reminded me that in almost all interpersonal relationships, no matter how positive they seem, there will likely be trauma and weird assumptions buried somewhere beneath them.

12 Tagging Animals in the Field

THE OTHER MAIN COMPONENTS for our ICARUS project were, of course, the tags on the ground. Without them, the antenna would be useless, and so we organized many trips to attach tags to animals in the field.

In one of our more ambitious tagging projects, we traveled to the Okambara Elephant Lodge in Namibia to attach ear tags to giraffes and other large African savanna mammals. This was very different from my experience tagging birds in Panama and Germany. Giraffes appear to live in slow motion when you observe them from afar, but once you get close to them, which you need to do when you are trying to immobilize them, you realize how fast they are. And how powerful. No wonder a lion never approaches a giraffe from behind for fear of a swift kick from its vicious hooves. A lion also doesn't approach from the front, where the swinging giraffe head with its two horns at the end of a long, powerful neck acts like a medieval battle-ax.

One easy way out is to have an expert veterinarian shoot a tranquilizer dart from a small helicopter, which is what we opted for. Then the ground team—us—following the giraffe carefully on a pickup truck has a minute or so to wrap a large rope around the front of the giraffe, which is now decelerating because of the tranquilizer. Once the rope is securely around the giraffe, four to six people are needed to slowly stop the massive giraffe body, and then some other people can wrap a rope around the hind legs of the towering muscle-mammal to slowly lower its body to the ground without hurting it. As soon as the neck and the head of the giraffe are on the ground, the team has two minutes before the giraffe regains full muscle function.

Our job in those two minutes is to attach ear tags to the animal—one click of our ear-tag pliers on each ear—to provide us with a lofty electronic view of the savanna. Nothing that moves through this African landscape escapes the view of a herd of giraffes. When all the heads are pointing in the same direction, this will alert us that something interesting is happening in *Lion King* country. And even better, the height of the giraffes allows the radio signals to travel directly into the air without bouncing around on trees or in the branches of bushes. Giraffes are simply the perfect biological lighthouses and watchtowers in the animal kingdom. Given a voice through our tags, giraffes are ideal spokes-animals for the savanna community.

Not all our animal tagging experiences were quite as physically demanding—or quite as successful. We had decided for one of our ICARUS projects that we wanted to tag endangered takins in Bhutan to learn more about their movements and see what might be done to protect them. Bhutan is perhaps the only country in the world where an intact natural ecosystem exists across the entire nation, from the lowlands, where Bhutan's major rivers flow into the Brahmaputra river basin, at an elevation of maybe 330 feet (100 m), all the way up to the highest peak at 25,000 feet (7,570 m; except it should be stated that a few years ago, China decided that the highest peak of Bhutan belongs to China...).

Bhutan has amazing people who live immersed in the natural world without intruding upon it. When you are there, exposed to the spiritual landscape of the high Himalayas and to the people who live there, you are enveloped in a sense of unity with everything around you. Humans see themselves as part of nature in this kingdom and they do their best to live their lives this way. Traditional values have served Bhutan well. None of the latest developments of Western or Eastern modern culture are immediately available in Bhutan, largely because the country is poor. And yet, it is perhaps the happiest country in the world because the government's first and foremost obligation is to make its people happy. Bhutan has a Gross National Happiness Commission, an institution every country should have.

Bhutan was interesting for us because animals migrate up and down the mountains across the kingdom. Within a distance of maybe 20 or 25 miles (30 or 40 km), Bhutanese animals can find any habitat, from the highest Himalayan meadows down to the evergreen subtropical rainforests.

A bird can cover that distance in maybe an hour. It would take a mammal much longer because the hills are steep and the forest is dense, as a primary forest should be.

This was exactly the kind of animal movement we wanted to track in our takin study. Takins are unbelievable creatures that look like a mythological cross between a chamois and a goat. They are not particularly large, but they are formidable and know how to defend themselves with their horns as deftly as lightweight boxing champions defend themselves with their gloves.

Takins stay low down in the winter. "Low" in Bhutan means elevations of about 8,200 feet (2,500 m). But the lowlands are places where they might interact with major predators, such as tigers or leopards. Tigers favor the lowlands because food is abundant—there are populations of gaur, for instance. This wild relative of the domestic cow likes to live in large groups like African buffaloes and is not an animal to be trifled with.

Sometimes tigers climb higher to hunt for deer and other mountain species that are easier to subdue than a huge 2-ton gaur. And sometimes tigers catch and kill some of the domestic cattle that are traditionally herded up to alpine meadows in the summer. Not only does the government pay a farmer if a cow is killed by a tiger, the farmer gets an additional benefit—the meat from the dead cow can be used for human consumption. Bhutanese Buddhists do not kill animals for food, but if a tiger kills a cow, the cow is considered to have died of natural causes and thus the meat can be shared between the tiger and the farmers.

When tigers climb up to these alpine meadows at elevations of 11,500 feet (3,500 m) in summer, they might also cross paths with a takin. Our aim was to understand how the takins move up into even higher elevations in the summer to avoid the tigers. Only the snow leopards venture higher, and they are much smaller than tigers. For a takin, the snow leopards are still somewhat scary, especially for the young ones, but by far not as horrendously dangerous as a tiger. The takins are also drawn to higher elevations by promises of fresh grass for their calves.

No one quite understood which animals were at which elevations when, so we decided to test our ICARUS GPS and animal movement tags on a takin. The first step was to see how the tag worked on a captive takin. This would allow us to create a baseline for takin behavior on our computers. How do they usually behave? Do they run, climb, jump, feed, fight, and so on? Just outside the capital city, Thimphu, is a takin reserve with a beautiful group of these majestic animals, including one that had been bottle-fed by two Bhutanese keepers. We were promised that this takin was completely tame and the keepers could do anything and everything with it.

When we arrived with our Bhutanese colleagues, the takin was the picture of contentment as it hung out with its keepers. We stood outside the fence, where many Bhutanese visitors often stood with grass to feed the takins. We had brought along a beautiful leather collar with a small ICARUS tag. Sherub, our Bhutanese friend, waved to his friends the takin keepers. They walked over to us, and we discussed how best to attach the collar to the takin. No problem, they told us. The takin totally trusted them. I noticed that the takin was taking this all in. From its posture, it appeared to be a little more alert than it had been when we first arrived.

It didn't seem to like the fact that some of the new people were talking to its friends. When the keepers walked back toward the takin, it stiffened.

One of the keepers had the collar with him, but the takin did not seem to be bothered by the collar. Rather, the takin was observing the second keeper, who was perhaps walking differently, as if on a mission. When the two keepers tried to slide the collar around the neck of the takin, it put its head down, clearly indicating that things were getting serious. No, it was not okay with having a collar put around its neck. An important lesson learned. What the keepers had told us—that they could do anything and everything with their beloved takin—was indeed true, but only when the takin perceived that they genuinely had its best interests at heart. As soon as other people (us) appeared to interfere and tell the keepers they had to do something, the takin stopped agreeing with its friends.

We walked away, hoping that the takin would forget our presence and accept the collar. But it never did, which meant we could never study the body accelerations and thus (remotely, via the electronic tag) the behavior of takins in captivity. We didn't want to anesthetize the takin as this would surely have scared it or turned it against its friends, the keepers. In the end, it wasn't an issue because we could also infer the behavior of takins by measuring the movements of other, similar animals, such as cows or chamois. But the takin story made me think more about humans' relationships with animals—or perhaps more accurately, their relationships with us.

What I witnessed in the takin reserve in Bhutan suggested to me that the animal was judging our relationship with its friends and saw that the behavior of its friends changed after they interacted with us. I was told the takin continued to be totally amicable and friendly with the two keepers, but this one instance was apparently a bit traumatic for both keepers. Neither of them had realized when they went into the tag attachment trial that they had an unspoken relationship of trust with "their" takin. This relationship changed when the keepers sided with strangers instead of with their "baby." Although this is an isolated incident, it still tells us a lot about how animals interact with people. They seem to have a lot more mental and social capacity to understand humans than we generally give them credit for.

Getting to know the locals in the Galápagos in the early 1990s: the ever-playful rice rats (above) and a sociable sea lion (below).

Early days tracking radio-tagged songbirds by car in Illinois in the late 1900s: a young Bill Cochran releasing a tagged thrush (above); Princeton undergrad Jamie Mandel with our customized tracking vehicle (below).

Learning how to track whole eco-systems using ARTS, our automated radio telemetry system, in Panama in the early 2000s: orchid bee (above); woolly opossum (below).

The latest ICARUS insect tags will be similar in size to this radio tag on a dragonfly (above), and those for larger animals will be about half the size and weight of the radio tag on this straw-colored fruit bat (below).

Tags and the methods used to attach them need to be sturdy enough to withstand some abuse and streamlined enough to not interfere with the animal's behavior: yellow baboons (above); scarlet macaw (below).

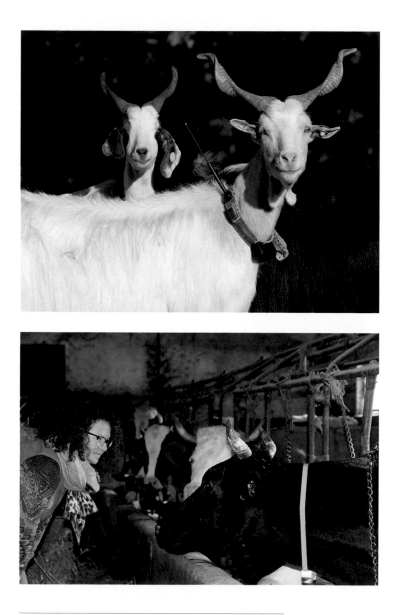

Close to home, data from tags on domestic animals can alert us to impending natural disasters: goats near Mount Etna (above); Uschi Müller with cows in the Abruzzo earthquake area, Italy (below).

Farther afield, ICARUS tags can gather information from animals that roam in remote landscapes: prayer flags in Bhutan (above); takin (below).

From the tiny to the titans, ICARUS tags can collect data every day from a wide variety of animals all over the globe to help protect wildlife, our planet, and us: Martin Wikelski with a white-crowned sparrow (above) and ear tagging a rhino (below).

13 Getting Closer to Launch

FEW PEOPLE CAN PROBABLY IMAGINE what it means after sixteen years of working toward a specific goal to be really, really close to implementation. In 2017, we shipped our computer to Baikonur in Kazakhstan, a mysterious spaceport that was one of the Soviet Union's most closely guarded secrets during the Cold War. It was chosen because its existence could largely be hidden from the West, and all the maps were misleading to confuse people about the location of the cosmodrome where the country's intercontinental ballistic missiles were stored.

I personally flew to Baikonur via Moscow to check the computer and the antenna before they were packed into the Soyuz capsule. I love how bold winter is in Moscow, with a deep cold you know will last for a long time. Landing in Baikonur with the temperature hovering around –13°F (–25°C) and the wind whipping across the steppe was a rude awakening to the harsh environment once you leave the relative comfort of the city. Baikonur is a fairly big town and apparently not totally dependent on the spaceport, but most of it felt like it was designed to support the space race. Payloads were still being sent into space, but the carnival atmosphere that must have reigned here in its heyday was mostly gone.

We were housed within the launch area on the site of what had been one of the larger launchpads. I had no clue that the entire launch area consisted of roughly fifteen individual launch sites, of which thirteen were now just ruins in the midst of a huge, open, flat expanse of steppe. At one time the Russian space program had been an important counterpart to the NASA program, but the launchpads had been abandoned when the Russian shuttle program was shut down and the space program was scaled back because of financial constraints after the fall of the Soviet Union. Nevertheless, the Russian space shuttle, Buran, is an impressive copy of the U.S. space shuttle. It is larger, and it felt like the Russians could have had an interesting program if money had continued to flow. We were told Buran had flown into low orbit on autopilot, although we didn't know if that was true or if it was just a story to make the Russian space engineers and the public feel good.

We were benefiting from the fact that there had been a Cold War because intercontinental ballistic missiles were the base for building the Soyuz rockets carrying cargo, mostly to the ISS. Because ballistic missiles had better hit their targets, Soyuz rockets are among the most reliable rockets in the world, which was comforting for us, as we had only one ICARUS antenna.

There was very little human activity going on, at least on the outside, because it was so cold. Even the hardy Bactrian camels we saw hanging about on the roadsides were hunkering down to escape the cold wind. Our rooms, built in the style of the 1970s, were, as is usual in Russian winters, completely and totally overheated. Apparently, there was no shortage of fuel and it was more convenient to run the heat continuously and open the window occasionally than to install thermostats. We had to fill out all kinds of forms, but because we were traveling with our Russian engineering colleagues who visited regularly to load equipment into the Soyuz capsules, they smoothed the way. However, we were not allowed to leave the hotel or walk up and down the road without security protection.

The next morning, we were brought to a huge building, the center part of which was completely off limits. Once inside the building we were not even allowed to go to the restroom alone. The rooms flanking the center were where food was prepared for the cosmonauts and where our ICARUS antenna was stored. It was really nice to see the dedicated Russian women cooking the best traditional food they could under the circumstances for cosmonauts young enough to be their children. I thought that if I were in space, I would feel incredibly well taken care of with all this food just like grandma made. They were proudly selling some of the meals that they prepared, vacuum-dried, at the little museum in town. We were excited to take a few home but disappointed to discover that after vacuum packing, the meals bore little resemblance to the excellent home-cooked food in Russia.

We were allowed to set up shop in a room that was completely empty except for a few benches along the walls. Our box containing the ICARUS antenna was eventually wheeled in. With its four wings (three receivers and one transmitter) folded, the rectangular antenna was roughly the size of a large fridge full of electronics and cables and plugs and protective material. The box looked intact, so that was a good sign. It hadn't fallen off the truck or the cargo lift. We carefully lifted off the top, well aware that this was our only antenna and if anything was deemed damaged or not perfectly functional, it would not be launched. So, no short circuits, no bent cables, no off-kilter screws. Everything had to be perfect. I couldn't have been more careful if it had been Michelangelo's *David* himself in there.

We tested all the functions of the antenna. Everything was looking good, when suddenly—why was there electrical current on this line but not on the other? Damn. The connectors had been built to allow currents to flow in one place but not another. This should never have happened. We had had design meeting after design meeting, and the circuits had even been engineered so the connectors could be switched and the antenna would still function. But the one mistake that should never have happened, had happened. The Russians, who had made the connectors, had connected the wrong cables to the wrong connector pins, but the mistake was hidden from anyone doing a visual check because the mix-up was inside the connector plug.

The antenna came with two connectors that looked identical. What this mistake meant was that the only way to prevent a massive disaster during the spacewalk to attach the antenna to the ISS was for the cosmonauts to be absolutely, 100 percent sure that out of two connectors that looked exactly alike, they picked the right one. The unique numbering on the connector sitting on the end of the cable coming from the ISS needed to be connected to our antenna's specific connector number in space. Could the cosmonauts even read this small number during a spacewalk? While on the ground it seemed doable, I had no clue how this would work 250 miles (400 km) above the Earth while orbiting at 4 miles (7 km) per second. We got into a huge debate. We had to involve everybody connected to our project—the German space agency, the German and Russian engineers, and the Russian space agency. Was it okay to launch the antenna? We finally got the okay.

And then another potential disaster loomed. The antenna was always, from the earliest onset of the mechanical design, meant to be launched in a horizontal position. You can imagine that sitting in a rocket on top of an explosion caused by burning 15 tons of solid rocket propellant to overcome gravity is not a pleasant place to be. Things shake, rattle, and roll inside a Soyuz capsule during takeoff. We had tested our antenna on a majestic machine, running through all the rattling frequencies that you could ever imagine could occur during the launch. And now here we were in Baikonur, after sixteen years of preparation, including ten years of design, with an antenna that needed to be launched in a horizontal position, that had been tested in a horizontal position... but there was only space in the Soyuz capsule in a vertical position. Oh. My. God!

This dramatic situation was not going to be resolved in a day. We halted deliberations in the late afternoon and postponed everything for the next day. What to do? All the design documentation stated that the antenna was not to be launched in a vertical position. We scrambled to discover the exact magnitude of the vibrations during the launch. Were they really going to be bad enough to destroy the antenna if it was launched in a vertical position? Some of the engineers said, "Yes, there's a high chance it's going to partially disintegrate." Others didn't think so. In the end, it had to be a team decision.

A lot was at stake. If the antenna broke on the way up, it would be discarded and we'd end up with a huge pile of expensive space junk. Apart from it being a complete disaster for the ICARUS project, such a failure would reflect extremely poorly on the German space agency and on SpaceTech, the German company that had built and assembled our ICARUS space hardware. And it would also be a major disaster for Roscosmos and RSC Energia, the human spaceflight company in Russia. But if we didn't launch the antenna now, we would most likely not be able to launch it for another two years and would have to completely redesign the antenna to fit it into a different space in another rocket carrying different cargo. This would need to be done by the Russian engineers. But they didn't have the funding or the time to do that.

In the end, it was up to me to try to convince people to either launch and run the risk, or play it safe and postpone the project for two years—or indefinitely. I also had to consider the legal implications. If I pushed for the launch and the antenna broke, was I going to be responsible for the failure of the entire project—$30 million on my bill? It was not an easy decision, as you can imagine. But there was one way to get help.

I turned to someone at the German space agency, who in his other life organized the annual carnival in Cologne, Europe's largest street festival. I guess you could compare the festivities to the Mardi Gras celebrations in New Orleans. But he was also the head of legal support at the German space agency, and at the time he was part of an international team working on a book of intergalactic law to formalize the legal aspects of space use. My contact, Bernhard Schmidt-Tedd, was also very sympathetic to our endeavor. I didn't know how to reach him immediately, so I called Uschi, the coordinator of the ICARUS project, because she usually solves all of these problems.

Half an hour later, Uschi knew Bernhard's current cell phone number and his whereabouts on holiday in the south of Spain, and she had already called him to prepare him for an emergency phone call on a difficult decision. Even though I had strict instructions from the security services in Baikonur not to step out of the hotel, I walked up to the road because that was the only place where I could get at least a tiny bit of cell phone connectivity. There was no connection I could use in the hotel that would be safe (not that my cell phone was necessarily safe...). I wanted some privacy as I tried to sort this all out.

I called Bernhard in Spain, where he had a nice dinner sitting in front of him, but he was happy to talk to me. *There is nothing better*, I thought, *than talking to an intergalactic lawyer from a freezing cold road that I'm not supposed to be standing on in the middle of nowhere in Kazakhstan.* Bernhard carefully explained to me the various ramifications of our options—and I decided we had to go for it. The next day I talked to everyone and finally got a unanimous decision for the launch. It was really hard work and everyone was extremely nervous, but it seemed to be the best option—or at least the least-damaging option—in the long run.

The next day we were coasting along nicely. We were shown the Baikonur launch site. We visited a local museum and the house where the first man in space, Gagarin, had lived. We got a tour of an abandoned launch site once used for the largest rocket ever built at that time. We were impressed by the concrete walkways over half a mile (1 km) long that the Russian engineers had built for people to escape through in case a rocket exploded, which had happened once before. Of all the cosmonauts whose photographs were on display at the museum—many of them coming from nations that do not exist anymore these days, the Soviet Union, East Germany, Czechoslovakia, Yugoslavia, to name but a few— only four had died in a space-related disaster. The priority of the space engineers had been to bring everyone back alive and well. It was all very impressive. Our job was done, and we went back to Germany excited about returning soon for the launch.

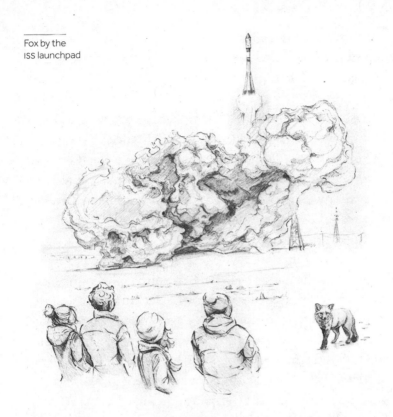

14 Finally, We Have Liftoff

ON OUR NEXT VISIT TO BAIKONUR, we were part of the official delegation from the German space agency participating in the launch of a Soyuz 2-1A rocket on February 13, 2018. There were no seats on the regular flights to Baikonur, so from Moscow we had to rent our own small plane and do it the VIP way. This time, we were not treated like part of the team of local engineers, but like foreigners who needed to be checked and watched carefully—or maybe the airport staff just wanted to show off and prove to everyone how efficient they were.

Instead of being put up in accommodations at the Baikonur space village, we were bused to a typical Soviet-era "space hotel." It was a place that got completely booked when cosmonaut launches happened, but on this occasion we were the only visitors—our group of sixteen in a hotel built to accommodate four hundred guests. It felt like we were in a horror movie. Only one short hallway was heated and lit. The rest of the hotel was dark and cold, and the staff was there only for us.

We had a couple of days before the launch and we filled our time seeing the tourist sights in Baikonur: going to the space museum, visiting the local market, checking out the street where all the cosmonauts since Gagarin had planted a tree, which was actually a very impressive collection of names and trees of different sizes going back fifty years. This visit was particularly moving for two of our space engineers who had grown up in former East Germany. They had been part of the Eastern Bloc of the Soviet Union at the time, and some of the early East German cosmonauts had also planted trees here. It was a weird reminder how entrenched political divides can be, yet how quickly things that seem to be set in stone can change.

While we were in the space museum, we came upon a reference to the U.S. moon landing in 1969. Our Russian counterpart, one of the head engineers on our project, asked how much of the story was true. Did the U.S. astronauts really land on the moon? Were parts of the story fabricated or was it all a lie? He openly questioned some technical aspects he found unbelievable. This got us thinking about how the moon landing had been reported in the Soviet Union and the Eastern Bloc, but also wondering whether we in the West should have asked more questions at the time.

Finally, the decisive morning arrived. We got up at four in the morning to watch the rocket being moved out of the hangar along the tracks that led to the launchpad. Communication was a little poor at this point and we didn't understand which launchpad was going to be used. But it was impressive to see everybody tuned in to getting the rocket moving slowly in the right direction—whatever direction that was. We knew our antenna was filling most of the cargo space at the tip of the rocket, and we felt like we owned this one. Hundreds of people were now working to launch the key component for the Internet of Animals. They were all just doing their jobs, but for us this was a transformational moment. If they did their jobs right, ICARUS would soon take off—hopefully with better luck than its namesake.

We watched the red taillights of the train disappear in the darkness of the steppe. We took the bus to the launchpad, driving by the camels feeding on the last bits of freeze-dried winter grass piled up in areas where the wind was strongest. The launchpad was not the one used by Gagarin, as we had expected, but another one. Why, we didn't know. Nobody offered any explanations. We were only told what not to do and where not to go. After we got off the bus, we all tried to get as close as we could, and we ended up some 100 yards away. It was dark and freezing cold.

As the supporting arm raised the rocket into a vertical position, launchpad security indicated it was time to move off. We walked back along the railway tracks and stepped back onto the bus, expecting to be driven to an observation platform. Except the bus drove slightly downhill and made a right turn onto a gravel road covered with undisturbed snow. We had literally been driven into the middle of nowhere. Apparently, we had arrived at the place. We hopped out to admire the launchpad slightly uphill in the distance. We could detect some activity around the rocket. A school bus with about fifteen Kazakh schoolkids arrived. It was their special day, as well. We exchanged a few words in English and then suddenly the side arms supporting the rocket moved downward. This meant maybe another five minutes to launch.

We watched with our binoculars, waiting for somebody to broadcast the countdown. But there was no countdown. This is an American invention, apparently. The rocket was now standing alone in all its majesty waiting for the signal to lift off. We all waited—and waited, and waited... Maybe we had misunderstood the time when the rocket was to be launched. Suddenly a white puff came from the engine. Was that a good sign or a bad sign? Fire trucks rushed up to the launchpad. So, maybe a bad sign. It was not getting any warmer in the freezing wind. After what seemed like an eternity but was only about fifteen minutes, the bus driver started the engine, indicating that we had better move. We tried to find out what was going on, but only learned hours later that the rocket launch was not going to happen that day or the next but maybe they would try again in two days' time. Super disappointing, but at least the rocket was safe and sound, it hadn't exploded, and our antenna was still in its tip. More time to explore the surreal, Star Wars–like atmosphere of an outpost town on a planet that doesn't exist, somewhere at the end of the universe.

Two days later, the same procedure. We were bused out to our very specific spot in the middle of nowhere, where we could see the faint impression of our bus tire tracks from two days before. All kinds of thoughts went through my head: *What if the rocket still doesn't work? Why did it not launch? Is anything wrong?* We hadn't been told anything. We just knew that supposedly there was no major problem. It was something minor. What issue that was, we had no clue. We stood there again and the fire trucks rushed down the hill again, away from the launchpad. And out of the blue, out of nowhere, a fox appeared. It walked between us and the rocket launch site. It either didn't notice us or was ignoring us, but for us it was like the fox that visited Antoine de Saint-Exupéry's little prince when he was grounded with his plane in a foreign land.

And suddenly, unexpectedly on this beautiful crystal-clear morning, the rocket started to rumble and shake, and a few seconds later we felt it in our guts. It truly was like a massive bomb exploding. The rocket started to inch upward. Clouds of snow flew around the rocket like a nebula. Then everything happened really quickly. The rocket gained altitude, we saw more and more fire from the engine, and the spectacle was over.

We danced, cried, and hugged one another. All the stress of nearly twenty years fell away. For that moment, we forgot how much lay ahead. The rocket had to reach the space station. The antenna needed to survive the launch vibrations. Everything had to fit together in the space station. The cosmonauts needed to fix the antenna to the outside of the ISS and connect it to the computer. And then we needed to see if everything actually worked. But at that moment, we didn't care. We no longer felt the cold. We danced in the snow. We forgot that for the bus driver and the workers, this was just another day. Some of them had probably done this hundreds of times, but we felt as Armstrong must have felt when he took his first step on the moon. Only for us it might have been a case of a big step for us and a small step for humankind.

Five months later we were back in the ground control center of Roscosmos in Moscow. The rocket had arrived safely at the ISS and the cosmonauts were ready to do the spacewalk. We needed to be in Moscow because officially we still owned the antenna. If something failed, it was our responsibility. We had to be there to make on-the-spot decisions—and perhaps be blamed if things went wrong.

We were denied entry to the real control room, which was reserved for the technical ground control staff. We were told that this was a complicated spacewalk. So far this was supposedly the largest instrument the cosmonauts had attached to the outside of the space station on their own. Our concern was the cables. We had seen how many cables were dangling from the outside of the ISS model in that cavernous hall where the old woman made a note of everything in her book. We were also worried about the cosmonauts getting the right connector to exactly the right plug. They had only one chance to get it right. As at the launch in Baikonur, there was no countdown, no information about what was going to happen next. We sat in the back room of the ground control center and ran NASA YouTube on our German cell phones. It ate up all the data cards we had, but we didn't care. It was our only way to understand what was going on inside the control room next door.

It dawned on us how difficult the long spacewalk was going to be. You don't just get up in the morning and go for a stroll. You go outside the ISS and then you have only forty-five minutes of light before you are on the other side of planet Earth, and it is once again pitch dark. You do the big stuff using the light and the small adjustments using your headlamps.

Overall, everything went well except the spacewalk took three hours longer than we expected—roughly seven hours, mostly for the ICARUS antenna alone, instead of the projected four hours. There was one tense moment when our engineers jumped up from their seats when "their" antenna dangled from its cable and flipped around and hit the outside of the ISS pretty heavily. On the ground, our engineers had only been allowed to touch the antenna when their construction suits were grounded so that no electric spark could damage a circuit, and the antenna had been buffered by a massive layer of foam. Now, their baby was being thrown around in space and, in their minds, being handled far too roughly. But it all worked. Except for the cable plug.

Our hearts sank when one cosmonaut asked the ground control station: "Is this the right plug? I can't find the other one." But again, for them this could have been part of their regular work on board. Then the cosmonauts found the other plug, connected the cables, and after seven hours the spacewalk was done. When they got back inside, the cosmonauts switched on the onboard computer, booted it up, and ... the system worked. This was the moment when we once again broke down in tears. We had made it to the next major milestone in our quest to talk to animals around the globe.

15 The Rocky Road of Tag Development

WHILE EVERYBODY KEPT THINKING that the satellite was the most innovative and difficult piece of the puzzle for ICARUS, the most challenging piece was actually figuring out how best to connect the receiver in space with the transmitter, that is to say the tag, on the ground. If you wanted to have a small and powerful transmitter, a true wearable for wildlife, you needed to design the receiver accordingly, and vice versa. The communication between the two would determine how much information could be sent and received and how often these transmissions could happen. But most importantly, the tags had to work for the animals that would be wearing them. Any animal wearing a tag should not be hampered in any way as it lives its life. An ICARUS tag should feel like nothing more than a large tick or scar tissue, something the animal would forget it was wearing. This was always the trickiest part of the ICARUS technology to develop, partly because of funding issues.

Space agencies have deep pockets and can, if they want to, provide adequate funding for space developments, but they generally don't provide any funding for developments on the ground. They tell you it's none of their business, and yet without the developments on the ground, ICARUS would be lost in space. The only way the development of miniaturized and versatile wearables for wildlife happened was thanks to funds from the Max Planck Society, where various vice presidents had the vision to support our high-risk endeavor.

Designing wearables for wildlife—even though this expression did not exist two decades ago when we started to conceptualize ICARUS—was a goal from the very beginning of radio telemetry for animals. The first devices were lightweight radio collars designed for mammals such as grizzly bears and elk. Collars were tried for parrots, but the best place to attach a tag to a bird would be on its belly, which is the closest place on its body to its center of gravity. As there's no good way to attach a tag on a bird's belly, the next best option is to attach a tag to its back.

Bill Cochran's old friend Arlo Raim was one of the most innovative pioneers in this field. During the early days tracking Swainson's and hermit thrushes, Arlo took the tiny one-twentieth of an ounce (1.5 g) tags Bill had developed and experimented with cutting a few feathers on the bird's back, applying a few tiny drops of eyelash adhesive, and attaching the tag to the feather stumps. A tag attached using this method would not interfere with the bird's

thermoregulation as it could still use its feathers to protect itself against heat and cold. Furthermore, the feathers
that were not trimmed lay smoothly over the tag, ensur-
ing that the tag barely altered the aerodynamics of the bird.
Most importantly, after the bird disappeared to continue its
migration—usually after a week or two—the adhesive gave
way and the tags fell off (as confirmed in aviary trials). And,
no matter what happened to the glue and the tag, the bird
would eventually shed its feathers, including the ones cut
down to stumps, and grow new ones. It was a perfect system
for a short-term attachment. Unfortunately, Arlo Raim later
died while doing what he loved most. In the flatlands of the
American Midwest, elevated railroad tracks were ideal places
to get good radio reception. Raim was run over by a freight
train after he had fallen asleep on an elevated track while
following northern cardinals south of Chicago.

When Bill and I followed individual songbirds during their
overnight migratory flights, we would capture and weigh
birds one morning and recapture them—a most formidable
endeavor—and reweigh them the next. We started following the thrushes shortly after sunset in Urbana-Champaign,
central Illinois, and let them tell us where to drive throughout the night. We tracked them as they flew to northern
Wisconsin, Minnesota, Indiana, Iowa, or Michigan. Thanks
to information we decoded from their tags, we learned from
these studies how much energy these magnificent little mara-
thon migrators burn up during their nocturnal flights. But
for me, these data were not the most exciting part of our
studies. What impressed me most was that the birds were
adjusting their own body weight to include the weight of
the tiny transmitter.

We would catch a Swainson's thrush in the south woods of Urbana-Champaign in the morning. Let's say it weighed about 1¼ ounces (35 g). Then we attached the little radio transmitter, five-hundredths of an ounce (1.5 g), and let the bird go. It ate throughout the day and fattened up by several grams by evening. At sunset the bird took off into the night sky and we took off too, racing along the Midwestern country roads to keep up with the aerial artist. In the morning, shortly after the thrush landed, we recaptured the very same bird, hundreds of miles north, and weighed it again. To my incredible surprise, the thrush would still weigh 1¼ ounces (35 g), but now including the transmitter weight. How could this happen?

Somehow the thrushes sensed what their optimal weight should be, included the tag in their calculations, and adjusted their body weight accordingly, both when they took off and when they landed. Apparently, each bird had individual optimal takeoff and landing weights, which now included the tag weight. Then we discovered that these birds adjusted their weight and balance on an even finer scale. By chance we attached some tags a little closer to their head and some a little closer to their tail. This ever-so-slight variation in placement was just a result of the natural variation in how we held the bird while we attached the tag. What we discovered was that the subcutaneous fat that birds have during migration was adjusted to accommodate the location of the tag.

For those who don't know about songbird fat distribution, it is a bit different from that of older humans (males at least), who may display Michelin Man–type rolls. Birds store essential energy for migratory flights as fat under the skin on their belly. However, this fat is not equally distributed but concentrated between the throat and breast at the front and between the cloaca and breast at the back. The breast in between these two areas is all muscle. If we by chance attached the telemetry tag a little farther forward, the bird reduced the fat above its breast, while slightly increasing the fat below its breast, and vice versa. These adjustments almost exactly resembled the fine-scale trimming pilots do so they can keep their plane at a perfect angle for their flight. We interpreted this to mean that the birds were not too bothered by the tags. They merely adjusted their flight positioning by carefully distributing and readjusting their mobile weight—in other words, their stores of fat.

There was a potential downside for the thrushes. A bird's fat is energy insurance, and more fat allows it to fly a greater distance during an overnight migration. A bird doesn't use all its fat on regular days, but on a really bad travel day, it might dip into its additional fat reserves, some of which would not be available if it was outfitted with a radio tag. We realized that even if we worked hard to make our tags low impact, we would always bother the birds a little bit and give them a little bit of a handicap. However, we sincerely felt that the long-term benefits would outweigh the slightly greater risks tagged birds faced, and we never had any indication that our tagged birds had lower seasonal return or survival rates than their untagged companions. Humans need to fix many problems in the Anthropocene, and we hoped our tags would help us do this while keeping impacts to wildlife to a bare minimum.

The tiny tags that we had developed and built with Bill were absolute pieces of art, the best that he could ever get out of a Sputnik-type transmitter. Their most powerful features were the incredible longevity of their small batteries and the huge distance their signals could travel. And Bill had based all this on old-style radio signals. Another ingenious twist to these radio tags was that instead of relying on short beeps, some of Bill's tags transmitted signals as a continuous wave. This drastically reduced the lifetime of the tags, but it meant we could pack other bits of information into the wave, such as air pressure, which told us the exact altitude of the birds all the time. We also recorded how the tag's antenna swung as the bird flew, which allowed us to count how often and how fast it beat its wings, which in turn gave us a way to quantify how much energy it was using. Other sensors we built into the tags measured the bird's temperature and electrical charges on the bird's skin. These tags then even told us the bird's heart rate plus the rate at which it was breathing. Based on a simple, Sputnik-type tag that Bill had modified into a continuous radio transmitter, we were getting a huge amount of information from the birds in flight.

But all of this was possible only when we were close to the birds and constantly following them in our cars. It pained us to admit that although we could happily continue to do exciting work with the telemetry systems we had developed, we needed to know what the birds were doing when we were not with them. When we started to develop ICARUS, we also realized we needed sensors that could measure and evaluate the data we were gathering from each tagged animal. To accommodate the increase in data that would have to be analyzed, we had to abandon our simple, superb analog system and go digital. Abandoning analog radio telemetry was both the most difficult and the most important development for the ICARUS tags and system.

In fact, this move to go digital is also what distinguishes ICARUS, our novel satellite communication system, from the analog Argos system that was available back then and is still running as a solid, dependable system today. Argos, a French and American satellite system, had been established in the late 1970s to read information from meteorological buoys around the world from space. Most of the buoys were in the vastness of the oceans. Later, the "buoys" could be built smaller and smaller, and eventually they were so small, they could be used as large animal tags. Initially, these animal tags on the ground were fairly inaccurate, sometimes being off by hundreds of miles. In addition, the signals from the Argos transmitters can get mixed up with other radio noises or communications devices such as taxi or boat radios. Thus, in areas around the world where there is a lot of this kind of radio traffic—the Mediterranean or Europe or parts of China, for instance—Argos is largely unusable. But it is a great system for traditional analog tracking of animals in more remote areas around the world. And so, it was not the best system for us as we worked to develop the smallest, most versatile, and cheapest wearables for wildlife globally.

Thankfully the rise of PDAs (personal digital assistants) in the 2000s and later the development of smartphones raised the same issues, and the digital revolution was upon us. The consumer industry developed systems to measure and analyze the position of a smartphone in three dimensions to know, for example, if it is lying on a table or if you have it up to your ear. Smartphones also routinely include temperature sensors to determine if the device is getting too hot or too cold. Magnetometers determine the direction you are facing when you use a map program. And your GPS sensors tell you approximately how high above sea level you are. Additional air pressure sensors can tell you the exact height in a tree or building, and a lot more combinations of sensors are available. For the ICARUS tags, we had lengthy discussions about which sensors to include, because we wanted to capture the most important information from the animals and their environment, but we couldn't have it all because that would consume too much power.

On the ICARUS tags, we had very little available energy and we needed most of it to send data over distances of up to 465 miles (750 km) from the ground into space. We decided to include sensors that measured acceleration and magnetic fields in three dimensions. We also opted for temperature, humidity, and altitude sensors that would allow us to tell whether a bird was in a tree or on the ground and what the weather was like in the exact location where the bird was flying. We knew this was a field that would develop rapidly and that algorithms to evaluate those data "on board" the animal would soon be available. The chip in each ICARUS tag needed to have enough capacity to run these algorithms when they were developed so we could review the data and program the chip to transmit only those pieces of condensed information that were important for us to interpret the bird's behavior and environment. All these components needed to be in a circuit board that weighed four-hundredths of an ounce (1 g) or less, because the other components of the tags were actually pretty heavy.

One of the heavier components was the GPS antenna. GPS is not, as most people think, a way to tell everybody where you are. The only thing a GPS can do is tell you where you are. There's no two-way communication in GPS systems. They can only receive information, and the receiver has to compare waves coming in from various satellites some 19,000 miles (30,000 km) away and make complicated calculations to give you your exact position, all of which uses up a lot of energy. Thankfully, you can buy the tiny electronic chips that do all this cheaply, but if you want to use them for animals, you still have to find the best ones with the lowest energy consumption and integrate them into your tags. And size matters. The GPS receiver chips with the lowest energy consumption often use the entire surface of your cell phone as a receiving antenna. In a little bird tag, all of this functionality has to be reduced to the size of a pinhead, which is one of the major challenges.

Once you have set up your GPS receiving antenna, you need a second antenna to transmit signals to the satellite. This requires an external omnidirectional antenna that will scan the visible sky, because that is where your satellite will be. It will just be up there somewhere and the transmitting antenna on the tag needs to find it, no matter where the animal is or how it is positioned relative to the satellite receiver. Unfortunately, omnidirectional antennas are always inefficient because they are not pinpointing a single spot, as a telescope does when it is trained on a single galaxy, but sorting through all the noise coming in from space. These transmitting antennas are usually in the form of a whipping wire that can send signals in almost any direction. A whip antenna for the ICARUS radio frequency needs to be roughly 6 inches (15 cm) long, whereas a slightly different style of antenna for the same system, a patch antenna, can be as small as 2⅜ inches (6 cm) in diameter. In short, lots of trade-offs need to be considered to design the best and most appropriate tag for different types of animals.

And now comes the really sad part—the battery. The battery is always your most important friend and your worst enemy, almost like your water bottle. It's the heaviest thing you carry on your hike, and you deplete it when you are most thirsty. The only way to get a refill for an animal tag is with solar power. (We are working on energy-harvesting devices like the movement mechanisms in expensive Swiss watches that run forever if you move your arm, but they are much less efficient than solar panels.) The main problem with solar is... it needs sun. Batteries already have a hard time functioning in cold temperatures, but imagine the pitiful amount of solar power you get from a tag in the northern hemisphere in winter on a bird that loves to hang out in dense vegetation. It is almost zilch. Compare this with what you have available when you study a tropical seabird with almost constant full sunlight hitting your solar panel. With this kind of power, you can happily determine the bird's GPS position multiple times a day, you can get the chip to do elaborate onboard calculations, you can write your data to your onboard storage. In short, you can make a series of electronic somersaults and still have enough charge to bounce in all kinds of directions.

With this massive variation in energy availability, programmers of animal tags have to think like military planners keeping all options, all fallback solutions, and all off-the-wall eventualities in mind. With a blackbird in the winter in the north, you have to make sure that the tag has just enough energy to transmit a short message every week or two while keeping the main sensing functions and the tag's memory ticking over. With a sooty tern soaring across the equatorial ocean, you are in energy heaven and you can—and indeed, should—use your energy-expensive GPS receiver a lot so the battery doesn't overflow with the incoming surplus of solar energy.

But there is more to a wearable for wildlife than how it functions. You also have to protect the components from the environment—and from the tag's owner and its friends and family. A rainforest is as continuously moist as your bathroom after a long, hot shower. A desert is as hot as a sauna. Winters can be colder than your freezer for extended periods of time. And I have not even begun to talk about the deepest oceans, the highest mountains, or underground tunnels. On top of all this, animals—like some people—often just don't take very good care of their electronic gadgets. Animals with huge beaks love to pick at things, young baboons like to chew on the antenna of mom's tag, the bone-crushing jaws of lions and hyenas can make short work of just about anything, and birds, as we discovered with our blackbirds, sometimes kink or break antennas when they preen. Luckily, in the case of the blackbirds, we found that nickel titanium alloy—used, for example, to make braces in human dentistry—works really well.

Finding a way to put all these electronic components into a small, ergonomic, sturdy package is an art. On a small circuit board, you have a lot of interference. The waves emitted from each of the dozens or hundreds of electronic pieces may interfere with each other, so you have to adjust all the components in relation to each other so that doesn't happen. The housing has to be strong but super lightweight. It also has to have attachments for a harness, a backpack, an ear tag pin, or other means of attaching the tag to the animal, and it has to be filled with glue or epoxy to keep the parts together and the electronics safe while allowing for a few small openings to the outside if you want to measure humidity or air pressure.

My friend Roland Kays, who works on mammals a lot, likes to test tags by grabbing them forcefully, trying to rip them apart, twisting them, hitting them against a rock, and biting them in all their most vulnerable places. If a tag survives this treatment, it might work as a wearable for wildlife. Needless to say, most prototypes and many commercially available tags do not survive Roland's stress test. And there's one even more important aspect. We didn't want to produce just one ICARUS tag, like a *Mona Lisa*, one piece of art that is the best of its kind. We wanted to produce a thousand, ten thousand, a hundred thousand *Mona Lisa*s.

And how do you attach wearables for wildlife to an astounding array of different types of animals? For now, let's just consider birds. We have studied everything from bee hummingbirds in Cuba weighing barely seven-hundredths of an ounce (2 g) to Himalayan griffons with a wingspan of up to 10 feet (3 m). We tested various methods for attaching tags to larger birds, and after many, many trials, the best material we found for a harness was a type of Teflon used by the military for parachutes. Military-grade Teflon is not degraded by the sun and is almost impossible to cut with a knife.

The next issue was how to fit a harness to a bird so it fit snugly but didn't interfere with the bird's lifestyle. Based on the advice and ingenuity of people from Mongolia to Russia, China, India, South America, North America, Africa, and Europe, our solution for larger birds was a harness lined with a flexible rubber structure. This harness smoothly adapted to the contours of the bird as it flexed and extended its muscles while constantly performing the activities it needed to perform to survive.

As a global community, we learned from trials and mis-
haps how to make the best possible system for a wide variety
of birds. Perhaps the longest and most intense trials were
done by my colleague and friend Jesko Partecke while studying his beloved Eurasian blackbirds. He and his team have
spent years figuring out what the perfect blackbird wearable
should look like and how it should be attached. What Jesko
discovered was that it's best not to use a body harness for
a blackbird. The most important reason is that these birds
almost double their weight in preparation for migration season. Thus, for blackbirds and similar species, the best system
looks almost like underwear, a kind of elastic G-string onto
which the ICARUS tag is attached. Blackbirds, and perhaps
all birds that have long legs and like to run on the ground,
have very muscular thighs and are not bothered by this kind
of a harness. Jesko devised a way to manufacture hundreds
of G-strings with ICARUS tags. Using this system, it takes
Jesko less than a minute to tag a blackbird.

Another important question was whether our tags would
in any way influence the birds' reproductive activities. Would
a wearable make a bird more attractive, either on the male
or the female side, or would it—in the worst-case scenario—
interfere with the act of copulation itself? You may already
know that when birds copulate, the male usually sits on top
of the female and presses his cloaca against hers. Could a tag
attached to the back of a male or a female bother either or
both of them enough to cause a problem? Thankfully, we have
seen no signs that it does, and this was a great relief for us.

We had a couple of final issues to consider. Would a
blackbird with a tag on its rump be able to receive a good
GPS signal or would its body get in the way? Or to put it

another way, how should we mount the GPS antenna so the tag would receive the GPS signals in the best and fastest way possible? As it turned out, the receiving antenna was horizontal enough to pick up the signals quickly as the satellite passed overhead. Now we had to consider how well the all-important transmitting antenna would work. We were concerned because the long 6-inch (15 cm) antennas we needed for transmissions to the satellite would stick out beyond the body of the blackbird and even a little bit beyond the tail feathers. If the antennas kept hitting the ground, that might interfere with data transmission. In the end, the performance of our ICARUS antenna on the blackbirds was good enough for all the tags to transmit their data to our receiver on the ISS, and we put our worries aside.

WE HAD A GOOD GROUNDING in developing tags for birds, thanks to all the work we had done with Bill Cochran studying songbird migration in the Midwest. But when we started our research on Barro Colorado Island in Panama to study species interactions in a larger ecosystem, we needed new tag types and new attachment types, particularly for mammals that so far had only been outfitted with collars. Our fallback was to go back to Bill.

"What do you think are good attachment techniques for mammals?" we asked. "Are there any better ones than the ones we know so far? Better than the neck collars most people use?"

Bill said, "Let me call Judy. She lives just down the road and was once married to a tropical mammal researcher. I worked with her husband in Panama. Sloth tags, that kind of thing."

This was a side of Bill that I hadn't known before—I thought he had only worked on birds and mammals in North

America. We called Judy, she was home, and we walked over. Bill and Judy were clearly good friends and Judy said, "Why don't you look around in my attic? All the old boxes from Panama are there. Maybe you'll find something interesting." We went up. What a treasure trove it was! Perhaps the single most exciting find was a stash of huge harnesses. These harnesses were made of a white nylon material 3 inches (8 cm) wide that was stitched together in an elaborate, careful, almost artistic fashion. We couldn't figure out what animal they were for. Perhaps a hippo? But not in Panama... Or a small forest elephant in western Africa or Indonesia? But again, not in Panama. We brought one down from the attic to show Bill and Judy, who started to laugh. "Oh, yes," Judy said. "Those were the prototype harnesses for manatees in the Panama Canal. The idea was to understand how manatees swam the length of the canal, how they interacted with the boats, whether they maybe even snuck into the locks with the shipping traffic." The first tests were successful, but in the end the study was not carried out and all there was to show for it were these amazing harnesses.

For our ARTS telemetry system in Panama, we started out using traditional-style radio collars for ocelots, agoutis, coatis, monkeys, and other terrestrial species. Those were very good and helpful devices but still not what we wanted to end up with. They often felt a bit too bulky, too clunky— simply not wearable enough. Most importantly, radio collars sized for animals when they are young become too tight as the animal grows, and radio collars sized for the animal as an adult would slip off its head when it was young. Radio collars are not ideal if you want to track animals as they are

growing up, striking out on their own, exploring, and finding new territories, which is what we wanted to do.

Imagine your own life. You may have gone to various
schools and worked in various places, but later, like most
people, you probably settled down in one place. Your daily
and seasonal movements became pretty much standard. You
might leave your home in the morning to hit the bakery
or coffee shop, travel by train or car or bike to work, stay
there all day, go to the gym in the evening, and then head
back home. You might do some jogging, gardening, or other
activities and then you hang out at home or go out to eat
somewhere, often in one of your favorite restaurants. You
settle into a groove.

Everyone, person or animal, has their own individual history. They have roamed the planet in different ways, they
have experienced different events at different locations, and
all this life experience is imprinted on them forever. If you
don't know about their individual experiences, you can't
really interpret their daily lives. And even more importantly,
you won't ever be able to predict how they might react if
something changes in their environment. And yet, these
environmental changes are ubiquitous in the daily lives of
animals. A new predator arrives in an area, the forest is cut
down, or a new forest grows up; it might be very dry in one
year or very wet in another. These changes might force them
to move to a slightly different area or to somewhere completely different. To predict any of those decisions—whether
an individual will stay, move, fight, or flee—you need to
know what the experience of these individual animals has
been. And this is exactly what we wanted to know in the animals we were studying. If we ever wanted to make progress

toward really predicting the decisions of individuals in the wild, we needed to get away from the traditional radio collars that only work for adult individuals.

But where on its body can you tag a mammal, ideally for much of its lifetime, without bothering it? A bracelet or an anklet might work for humans but neither is good for four-legged creatures. We do know from livestock that ears are a good place as long as the tags are small and lightweight and the antennas aren't too long. We discounted the kind of tags you sometimes see on cows grazing on a pasture. The number is usually written on a big flap hanging down from the point of attachment. But we quickly realized that mammal ears move so much that out in the wild a large flap would spin around until, eventually, the ear tag would widen the piercing so much that the entire tag would fall out. So, instead of having our tags hang down, we located them directly above the pin that goes through to attach them to the ear.

Our ear tags needed to be small and lightweight, not only because we were mounting them above the point of attachment, but also because we wanted them to be true wearables. An ear can't carry a lot of weight—and it shouldn't—because animals are constantly moving their ears to listen in various directions or to flick off flies that hang around their eyes or head, as many mammals such as buffaloes, zebras, and gazelles do.

Another important design feature for our novel ear tags was that the surface had to be transparent so we could use solar panels to generate energy. A small battery would be able to transmit only a limited amount of data over a limited amount of time, whereas an inbuilt solar panel could keep an ear tag going almost forever.

For our first ear tag prototypes, we used the regular cow ear tags but glued or tied the novel electronic ICARUS tags to them above the ear pin, so they would sit neatly as a true earring, not as a dangling ear flap. Most colleagues we talked to expected these trials of solar ear tags to be utter failures. But everything worked. The solar panel provided enough power, the special antennas worked for communication, and the central placement of the ear tags was key to their success.

The first generation of ear tags worked so well we were confident that we could eventually develop a system that would last for the lifetime of a mammal without interfering with its lifestyle. But as usual, the first 85 percent of the work is easy and the last 15 percent is tedious and difficult. After years of trials, we have what we feel is a perfect system for tags that is free for whoever would like to use it.

Needless to say, there were and still are many issues to resolve. We have to find the right positioning within the ear in each species, and we are grappling with the question of whether the tag should be positioned on the front of the ear or sit on the back of the ear. When facing to the back, the ear tag might get smashed against the hard horns of a mammal or perhaps against its hard skull bones. When facing to the front, the tag may or may not get enough sunlight onto its solar panel. In a forward-looking position, the tag may additionally bother the eyes of the individual whenever the animal waggles its ears in front of its eyes to get rid of the flies living off its tear fluid.

ALTHOUGH OUR TAGS are still a work in progress, we are proud of what we have achieved. We have developed a new way for animals to be outfitted with electronic tags that record not only the animal's GPS location, but also the temperature and humidity around it, the behavior of the individual, and potentially its interactions with other individuals of the same and other species. We are using our ear tags in the Amazon and in Africa, where we plan to help protect various species of endangered mammals, such as rhinos and giraffes, by giving them a way to talk to us. We have tested this reporting system with a number of wild species and it seems to work beautifully. For example, we detected African swine fever in wild boar in Germany within three hours of the animals getting infected. Like us, when animals get sick, they slow down, and whenever our ear tags detect that a wild boar starts moving its ears more slowly, we know it is likely ill. When a decrease in ear waggling occurs within a few hours, there is a high chance the boar is infected by African swine fever. Detecting such diseases early in wild animals closely related to farmed animals is exceedingly important to protect livestock from contagious diseases.

A similar system also tells us when wild dogs are being snared. My friend and colleague Louis van Schalkwyk has used a remote, tag-based activity reporting system to protect highly endangered African wild dogs. Louis saved enough wild dogs each year in Kruger National Park that the population has switched from decreasing to increasing. How wonderful is that? Electronic tags now enable wild dogs to report when they need help, and whenever one of his beloved wild animals calls in, Louis drives out into the national park at any time of day or night to free them from snares. Sometimes the wounds are so deep it is hard to believe the dog will survive, but its family rallies around to help it recover and heal.

In some sad instances, the tags also tell us immediately—within ten minutes or so—that an animal has died. When the tag records that the animal is completely still, it calls in and reports that a life has been lost. We feel this is an incredible advance over the original beeping transmitter tags that we used in our rainforest telemetry system in Panama. With these new tags, a manatee or an agouti could now tell us whether they are stressed by vessels or by an ocelot, or whether they are excited because they have found a great food source or because they are in immediate contact with a conspecific. In just two decades we went from a battery-powered collar with a beeping transmitter with a limited life to an intelligent smartphone-style ear tag powered by the sun that the animal could wear all its life. And this novel tag has onboard intelligence to detect what state the animal currently is in, which it communicates to researchers via a satellite connection. If necessary, we can also get the data via a handheld receiver, but that requires locating and getting close to the animal on the ground, which is not an easy thing to do and often impossible.

We now have a power supply and method of attachment capable of lasting a lifetime. However, when you build a wearable tag and figure out how to attach it to the animals, initially it might look wonderful, but after a few months of being on the ear of a rhino, the back of a bat, or the leg of a bird, things can change. The epoxy above the solar panel becomes gray, or the surface coating cracks, or the housing degrades under ultraviolet light, or the bottom of the housing touching the back of a bird may be too rough even though it feels smooth. In addition, the tag may be waterproof initially, but not over long periods of time. The antenna may rust. And so on, and on, and on, and on... Like so many other situations in life, you can't simply buy experience or speed up developments, because you basically only learn from the mistakes you make. Thankfully, making mistakes is easy, but the learning process still takes time.

Over and above this normal learning process, understanding what exactly went wrong in individual situations is really difficult. If a bird stops sending data from Inner Mongolia, you first have to locate the bird and then you have to observe it or recapture it before you can find out what happened. Then you need to fix your mistake in a way that doesn't affect any of the other tag components. You can't always implement the optimal solution, because it might be too expensive or too difficult to manufacture in the future. The tag needs to remain simple and inexpensive enough that it can, eventually, be mass produced so that we can scale up global animal tracking to hundreds of thousands of birds, animals, and insects.

In many ways this overview of the difficulties inherent in producing wearables for wildlife is the story of unsung heroes. There are many teams around the world that have worked incredibly hard to get these systems up and running. For the public and for most of our colleagues in biology and ecology, it might look as though we, as a global community, just came up with these designs, but all of us tested these systems for many years and experienced many painful and frustrating failures before we finally had prototypes that functioned in the way we wanted them to.

Sooty terns

16 All Systems Go—or Not

WITH THE ANTENNA ON THE ISS, we were finally poised to go live and go global. We had designed it to help all of us move forward into an age where instead of dismissing or exploiting animals, we were seeing them as partners and protectors of life on our incredible planet. We would supply the tags and people could design their own projects. We were standing by to say, "Get ready, get your projects prepared, get your permits ready, and we will send you the ICARUS tags to deploy." Having worked toward this goal with a huge team for almost twenty years, a positive result would mean the world for us—literally. If our communication scheme did what it was supposed to do, we could truly say we had gone from Sputnik to a global Internet of Animals in the blink of an eye.

But first we had given ourselves from September 2018 to February 2019 to test and optimize the system. The first step once the computer and antenna arrived at the ISS was to test the ground-to-space communication by sending computer-generated tag messages into space from large and powerful antennas on the ground. Then we planned to switch on the tiny ICARUS tags attached to animals and get ready to send their data over the ICARUS channel to the antenna on the ISS.

There were so many questions waiting for answers when that switch was made. Would the tags really know when to send the data to the computer on the ISS? For this to happen, the ISS had to send down a signal that arrived exactly at a specific second. Then the tag had only 1.5 seconds to respond and transmit its data. However, with the Doppler shift affecting wavelengths at an altitude of roughly 250 miles (400 km), deciphering data can be extremely difficult. Would the ICARUS tags really be able to deal with both the short time frame and the change in frequency of the radio waves? We could simulate both on the ground, but not really test the tags' response. You cannot move at 4 miles per second (7 km/s) on the ground or in the atmosphere—nobody can. The only way to do it was the hard way: launch the system, switch it on, and hope that all the calculations were correct.

In the middle of September, without any notice, the Russians decided to start the ICARUS onboard computer without telling us. We got the log of the procedure later. It showed that the computer had started to boot, but after a few short moments it had stopped. What had gone wrong? Our engineers devised a way to test whether the fan was not getting enough electricity or was burned out. With our fingers crossed that it was nothing major, we scheduled an onboard repair exercise. The Russians gave us one hour of the cosmonauts' time. They took the onboard computer out of its rack, fiddled around with the on/off switch, unplugged one of the fans, and tried again. It still did not work. There was nothing for it: the computer had to come down and the replacement computer had to go up. Thankfully—unlike the antenna, of which we had only one—we had built two onboard computers.

We prepared the ground computer for launch while the onboard one was being sent back down in a Soyuz capsule. The computer landed somewhere in Kazakhstan and was brought back to Baikonur, flown to Moscow, checked there— and no issue was found. Then it was sent to the Russian company in Saint Petersburg responsible for the electric power supply. They discovered that a 1.5-cent capacitor had burned out, perhaps owing to space radiation. This was good news, because it meant the replacement computer could safely use the same power supply in the ISS once it had been equipped with a tested new capacitor. This being Russia, we could not be absolutely certain that this was what had happened, but we accepted the explanation and moved on.

The replacement computer was brought to Moscow, tested, flown to Baikonur, tested again, and launched. The cosmonauts put it into the rack, connected it, and it worked. It was now December 6, 2018. Finally, we thought, we can start the ICARUS project. But the Russians told us that, before that could happen, they needed to know from the ICARUS engineering team all the details down to the codes and blueprints of how the system worked. Obviously, their aim was to understand the system so they could build one of their own.

Did we want to hand over all this information? How much information did it make sense to hand over? Eventually we wanted to make ICARUS public anyway, similar to the GSM (global system for mobile communications) or cell phone systems, because only then would other entities around the world potentially launch their own ICARUS systems to supplement and expand ours. There was a big debate within our engineering team. Two of the engineers did not want to share any information and left with a lot of internal, proprietary information it took us months to recover. Not only were we dealing with the Russians, who constantly came up with new demands, but we were also dealing with internal disruption and dissent. People who feel like underdogs often want to monopolize and control information to protect themselves from outside forces. This is perhaps why helping others advance can be the most incredible tool to keep the peace. Eventually, we resolved our internal issues. The two rogue engineers were gone, and we gave the Russians enough information to understand the principles of the system, most of which was public anyway because we had just published it in a scientific paper.

All the setbacks ate away at time. ICARUS eventually went live on March 20, 2020, at almost exactly the same time that the global coronavirus pandemic started. The irony was not lost on us. We were finally at a stage where we could work with the world, but the world had shut down and could not work with us. Our science teams seized the opportunity and used the anthropause—the pause in human activity around the globe during the shutdown period—to run repeated natural experiments to quantify the influence human activity has on wildlife around the globe. They discovered, among other things, that human activity influences the time of day when animals can be active. For example, when humans are out and about, many animals such as deer or foxes become active mostly at night, but they start to roam in the daytime when humans are gone. And human activities change the way animals use space. Animal movements are more restricted in areas with high human impact, probably because individual animals change their behaviors as a result of the physical limitations human activities impose upon them.

As the ISS continued to circle planet Earth, the cosmonauts worked as usual and the ICARUS antenna worked like a charm. The antenna scanned the globe for background electronic noise in the frequency range we were using, and we could start to send data from the ground up to the ISS. It all worked beautifully, except that we had to reprogram the software and firmware for all of our ICARUS tags because our rogue engineers had left with some knowledge that they did not want to give us. It was unbelievably disappointing that people could be that selfish. But perhaps one always has to count on that.

While the pandemic was ongoing, we prepared our first projects. One project looked at Eurasian blackbirds to find out why birds leave their homeland in the first place and why some blackbirds buck this trend. We wanted our tags to be less than one-thirtieth (3 percent) of the weight of the bird that was carrying them, which meant for this project we had to reduce the weight of our tags. Russian blackbirds could easily carry a tag weighing one-fifth of an ounce (5 g), but blackbirds in Spain were smaller. This was when disaster struck again. When the tags were reduced in weight, there were some last-minute engineering changes.

The bird harness was supposed to be fed through a small rubber tube on either side of the tag, but somebody somewhere in the process had decided that Teflon was a better material for the tube. Although this shouldn't have been a big deal, it turned out the housing of the tag could not be glued securely to a Teflon tube and water leaked into the tags. We did not catch this when we tested the tags on the roof of our institute, but only later, when the tags were on the birds. We now had bird tags that provided us with only a few data points before they stopped working. One time NASA built a spacecraft where one engineering group calculated in inches and the other one in centimeters. The spacecraft was launched on a trajectory to intentionally crash-land on Mars—and disappeared. We felt as devastated by our failure as the NASA team must have felt about theirs.

Our team rallied and continued. Our larger one-fifth-ounce (5 g) tags were generally doing fine, and we repaired this small mechanical problem in our smaller blackbird tags. We also had several other species where ICARUS was proving to be an excellent tracking system for large-scale and small-scale movements of animals; however, there were still a few things we had to learn the hard way.

We were testing the first ICARUS ear tags on large African animals such as rhinos. The long, slender antennas were incredibly flexible and lightweight and, we felt, less intrusive than a grass stem, but apparently they were tickling the rhinos' ears. We also realized that rhinos can flick their ears to produce forces up to 8 g, which is incredible. These acceleration forces make the long antennas act like a whip. And if you whip them back and forth too often, they break—even though the material is supposedly unbreakable.

We also saw that the design of the rhino ear tags was not ideal, because the edges of the tags collected mud. And we learned, again the hard way, that the mud baths rhinos take are incredibly sophisticated. They are designed to separate ticks—one of the most highly evolved parasites on planet Earth—from the rhinos' tough hide. When a rhino with an ear tag took a mud bath, drying mud accumulated around the base of the whip antenna, and if the base was not 100 percent secure and stable, the antenna snapped like the neck of a tick. We could not believe it, but there it was, another lesson from nature. It was not enough to design a system to communicate from ground to space, you also had to design hardware that would survive a roll in the mud with a rhino.

While we were making enormous progress on all of these issues, there were, once again, mechanical problems on the ISS. Between December 2020 and March 2021, the Russian life support system experienced massive problems and all non-essential systems were shut down. The situation was resolved eventually, but in the meantime our tags had been searching for the signal from the ISS every day because we had programmed them to keep searching if they hadn't heard from the ISS for a few days. Our aim had been that the tags would always continue to communicate with us. When the ICARUS system completely shut down because of the power outage, the tags went into an endless search loop and eventually used up all their energy and in some cases even damaged their batteries. This was a huge problem for many studies, such as the one with sooty terns in the Pacific, Indian, and Atlantic Oceans. It was also a problem for the cuckoos that were on their way from the southern African nation of Angola to return to their breeding grounds in Kamchatka in the far east of Russia. To avoid the loop, we reprogrammed our tags to expect problems and only search for the right orbit parameters of the ISS every few days instead of continuously after a gap in communication, but unfortunately, we could not reprogram any of the tags that had already been fitted to animals.

By March 21, 2021, a year after going live, the ICARUS system was finally working as it should. We were connecting with colleagues around the world and had tags on all kinds of animals. The appendix to this book outlines the many projects we were getting up and running at this time. We felt like this was the start of a real global observation system. In April 2021, despite the continuing global lockdown, we made every effort to travel the world, from South Africa to the Seychelles, to Cuba, to Hungary, or to Sicily to fit tags on a variety of species of animals. We also sent tags to many collaborators all around the world, especially in Russia.

From April 21 onward, we started to get amazing data. Every day we sat at our computers watching the data stream in. Our red-footed falcons were on their way from Angola back to Hungary; the Hudsonian godwits were making their nonstop flights from Chile across the Galápagos and Guatemala into Texas; the supposedly stationary black coucals, an African cuckoo, were migrating more than 620 miles

(1,000 km) from southern Tanzania to northern Democratic Republic of the Congo; and the Oriental cuckoos were traveling from the far-northeastern Russian island of Sakhalin across Japan to their wintering sites in Papua New Guinea. There were still more interruptions than we would have liked. We had to shut down the system anytime anything was delivered to the ISS and whenever there was any slight problem on board. Nevertheless, we were content with all the beautiful new data flowing in.

But soon we got word of the next potential disaster. After years of lying on the ground in Roscosmos's hangar in Moscow, Russia's Nauka module was about to be launched to the ISS. We had thought this would never happen, but we were wrong. We had known about the Nauka module since 2012. It had its own internal engines, so, in principle, once it was attached to the ISS, you could uncouple the other modules and build a new space station around it. Nauka had had lots of problems. We were supposed to have had our ICARUS antenna fixed to the outside of Nauka back in 2014, but then Nauka needed a complete overhaul on the ground and we were happy that our antenna would be going directly to the ISS. Whenever we visited Russia to check with the engineering team, there was Nauka, lying on the ground. We often crossed paths with engineers from the European Space Agency, who had to move the European robotic arm about every six weeks for eight years so that the arm would not seize up while lying around not doing anything.

On August 21, 2021, Nauka was affixed to the top of a huge rocket and finally delivered to the ISS. We had prepared our engineering system and communication modalities so that a huge module sitting next to our antenna would not interfere with ICARUS's communication systems. Or so we hoped. We could not know for sure until the module was up there. That's the problem when you are working with waves. They are notoriously bouncy. You can guesstimate what will happen if things change, but you cannot simulate and test for every variation.

Nauka arrived and ICARUS still worked like a charm. How beautiful. But wait! Data from the ICARUS antenna suggested that either our sensors had malfunctioned or somebody had switched the cables, because the only way to interpret what our sensors were telling us was that the ISS was now flying backward. Which could not be the case. But it turned out it was. Without telling anybody, at least not us, the Russians had turned the ISS around 180 degrees. And Russia continued to do this many times in the weeks that followed. The problem for us was that when the ISS spun, our antenna spun along with it, and it could no longer receive signals from the ICARUS tags.

It was unclear to us if this was a power play by the Russians, the result of a system failure, somebody just wanting to have a little bit of fun, or something else. Eventually the ISS calmed down, only now its longitudinal axis had settled at a different angle. Although we could have fixed this problem easily, we could not get the Russians to send up the correct parameters so we could adjust our receiver in space. Our tags have only 1.5 seconds to communicate with the ISS. For the tags to hit this short time window, we have to let our tags know when the ISS will be directly above them so they know exactly when the receiving window starts. Knowing the receiving window ensures that the tags send the information at the correct time. The computer inside the ISS waits for the tag information and then decodes the information it has been programmed to listen for.

This should have been easy to fix, but the Russians, it appeared, had put our system on the back burner while they were occupied with something else: a movie. The Russians were determined to be the first to have an ISS movie. Although this might have been important to them in many ways, for us it was a disaster because they spent a lot of time positioning the ISS in such a way that Earth could be seen perfectly from the lookout point of the ISS known as the cupola. Once the movie was finished, our receiving window was still different from what we had told our tags. Aaargh! We sincerely hoped their corny B movie would sink without a trace.

It was now October 21, 2021, seven months into the effective start of our global observation system. The tags were wasting precious energy desperately searching for the ISS because we could not get confirmation from Russia that we were allowed to change the information sent to the tags to keep them in the loop. How could the Russians not understand that this was essential? What was so difficult to understand about needing to fine-tune the antenna you use to communicate between the ISS and the tags when the tags have to relay information? We could not fathom what they were thinking. Either this was yet another ploy to get more information from us about the tags—their design, their production, their programming—or there was some other message they wanted to send by almost deliberately handicapping the system until it was practically useless.

We were in a pandemic lockdown, so we could not just fly to Russia. We tried everything through all the channels at our disposal to change those few software parameters telling the tags the X and Y of the ISS. At that point, Roscosmos told us we needed another contract signed by all kinds of people before this could be done. We got all the signatures from our end and couriered over a hard copy of the documents. We hoped everything would be resolved before the European holiday break in December 2021, but of course nothing was. Then came the Russian Orthodox holidays that lasted into late January 2022. Eventually, at the end of February 2022, only one last signature was missing.

17 Animals at Play

WE WERE POISED TO LEARN so much about and from the animal kingdom. The story of Hansi the stork is a wonderful example of one way animals in nature come up with new behavior: pure necessity. Poor Hansi must have been so desperate for food that he approached areas around humans where he thought there might be some, such as on the farm. Maybe he also smelled food, approached it, and wasn't chased away. So far, we don't know the details, we just know that there is not enough food for storks on pastures in the middle of northern Bavaria in November. There simply isn't any. That's why most other birds—those that eat rodents or insects—are gone. Except Hansi. He found another way of surviving. But animal innovations also happen in exactly the opposite situation: when resources are super abundant and food practically jumps into animals' mouths of its own accord. There are two instances when I realized that this might be the case.

One of these instances occurred in the Galápagos Islands. Over many years, I had probably spent a total of thirty-six months on these islands observing the behavior of marine iguanas. One year El Niño hit. All the food for the iguanas disappeared, and up to 90 percent of the individuals in one population died of starvation. The water in the upper layer of the ocean was too hot and nutrient deprived to support the growth of marine algae in the intertidal zone. Thus these vegetarian reptiles, which graze for algae when the tide retreats, were obviously doomed. But then, a year later, the population structure changed entirely. The boom time for marine iguanas came as the cold waters reappeared, welling up from the depths of the ocean, bringing with them nutrients that allowed algae to grow in such quantities that the iguanas that had survived could graze to their heart's content. Everything a marine iguana could ever desire was there: warm temperatures on land, no rain, and sunshine all day so the iguanas could quickly run into the intertidal zone to gorge themselves on their favorite food.

It was in the recovery period after El Niño that I first saw what to me was an unbelievable sight. These scaly, cold-blooded creatures started to play. The highlight for me was when an adolescent male iguana grabbed a feather off a blue-footed booby standing on the rocks in the midst of the iguana colony. One of the bird's tail feathers was hanging down, and it caught this young iguana's attention as it walked by. The iguana grabbed the feather and ran away with it, only to set it down onto the rocks and look around. Immediately, another adolescent dashed up, picked up the feather, and ran off with it, but again not far. Maybe just 7 feet (2 m). It then dropped the feather and looked at the first iguana. The first one ran up to the feather and tried to get it, but the second iguana picked it up shortly before his challenger arrived, turned around, ran another few feet, and dropped the feather again as it looked at the first iguana. Then a third marine iguana came from behind the second one, ran past it, grabbed the feather, ran a short distance, and sat down again. Then the first iguana ran up to the third one, grabbed the feather, and ran off. And so it continued for about ten minutes.

This is the first time I have described this behavior because most people and most of my colleagues would have a hard time taking me seriously. But it is how it happened. And although there are many definitions of play in animals, my definition is that I call it "play" when I see animals play. I know play when I see it because human play is fun, and I feel it is fun for the animals, as well. I can see animals having the same motivation as humans who play. And it's clearly play when there's no other apparent reason for doing what they are doing other than to have fun.

This was my introduction to unexpected animal play and to the innovative behaviors of animals when life is bountiful. In general, racing around and playing tag is a common theme for young animals just as it is for young humans. One of the features of life that is apparently important for individuals and populations is that when you have surplus energy, you can explore, you can innovate, and you can do so much more than merely survive.

In the other situation, a colleague was studying songbird endocrinology in the High Arctic. This might sound weird to most people. Why would you study the hormones of birds in the first place and, even more weird, go to the northern slopes of Alaska to do so? Well, there is a good reason. We need to understand how our own human hormonal constitution evolved, why our body has and uses testosterone, and how this hormone competes with corticosteroid stress hormones and other steroid hormones. This is what biologists do. They care about and study the behavior and physiology of wild animals in their natural surroundings in part to understand human biology. Not unlike behavioral scientists who track animals with tags, physiologists also try to learn from nature—and their work helps us understand why our hormones play games with us at times. Everybody who has had to deal with kids during puberty knows exactly what I'm talking about: hormones are not to be underestimated.

The year my friend was in the Arctic was a year unlike any other. Conditions were perfect. Songbirds were nesting all over the low tundra vegetation, where you can't really hide well as a songbird. There was also not a lot of wind, so the songbirds couldn't even hide their smell. The arctic foxes, which usually have a tough time surviving, were having the time of their lives. They only had to run a short distance to find the next bird nest, where they could eat the eggs or the chicks. The foxes were in heaven.

My colleague had to make detailed observations of the nests she hadn't yet lost to foxes. She had discovered that this season, the foxes, which had been afraid of her and had stayed far away from her in the past, were now observing her all the time instead of going out and finding nests on their own. Before I go on with this story, there is an important fact you should know. Biologists have claimed that only trained dogs understand what it means when you point at an object. When a hunter, for example, looks steadily in one direction and points with a finger, their dog understands that this is the direction in which something interesting is happening and off it runs to find the hunter's shot bird. That's the accepted wisdom, anyway, but as we have been learning, accepted wisdom isn't all it's cracked up to be.

It was very clear to my friend that the arctic foxes didn't know that only trained dogs can decipher what it means to point. When she set off to observe her nests, she took a meandering route to confuse the foxes, because it hadn't taken them long to learn that she was observing birds. When she was looking with her binoculars, the foxes also trained their sights in that direction. What my friend had to do as a countermeasure was to scan the tundra in every single direction to deceive the foxes. If my friend looked in one direction for long enough, the foxes would run in that direction, find the nest, and eat the chicks.

My house cat is just the same. A house cat is not supposed to have anything to do with the fancy pointing behavior reserved for specially trained dogs. If you point in a certain direction, a cat is supposed to look at you and not at where you are pointing. Many people claim this is an indication of the superiority of the dog-human relationship that allows a perfect understanding between humans and the dogs they train. Dogs understand that they have to look away from a finger that is pointing, not directly at the fingertip. And that they then have to run away from the human and not toward the finger. But I can tell you that my cat is better with pointing than any dog. My cat grew up in a home where the family put the cat into an old bathtub together with a guinea pig.

My cat grew up fighting for food with this guinea pig in the bathtub. Our family rescued her when she was a kitten and she has been a lovely member of our family ever since, but she has never given up her deep and strong desire for food. My cat knows exactly what pointing means. I got in the habit of hiding a piece of food, and pointing in that direction. At first, my cat just looked at me, meowing. But after showing her a few times and telling her, "You have to go where I point, not where I am," she caught on. And now, I hide food, I point, and Luna knows exactly where to go. Perhaps there are very few things animals can't do if they are motivated enough, and sometimes I wonder who is training whom.

One morning our friend in the Arctic went out again to observe her songbirds. She took her circuitous route to throw off the foxes and settled down to observe the parent birds as they flew around and interacted with their neighbors. An hour or two into her observations, out of the corner of her eye, my friend caught sight of an arctic fox approaching her. When the fox was about 65 feet (20 m) away from her, it stopped and sat down. What the fox did next took her completely by surprise. The fox picked up a small branch that had broken off a bush. It was about 6 inches (15 cm) long and narrow enough that the fox could carry it easily in its mouth. It was almost like the marine iguana in the Galápagos that grabbed the booby's tail feather, but the fox didn't pick up the stick to entice other foxes to play. Instead, it carried the branch to my friend and stopped a few feet away from her. The fox had been observing her for weeks, so it knew it could count on her not to make any sudden movements. I'm sure that the fox also knew that my friend didn't have a gun, only her binoculars. But can you imagine:

a wild arctic fox, picking up a little stick, running up to you, and dropping the stick in front of you, almost as if it wanted to say: "Play with me!"

My friend, who loves dogs, carefully picked up the little branch and threw it off at an angle away from the bird nest. And you probably can already imagine what came next. The fox ran after the stick, grabbing it while it was still in the air like a well-trained dog. It then turned around on the spot and brought the stick back, laying it down again in front of my friend. And she threw the stick again. Unfortunately, at this point the little songbird returned and my friend needed to go back to her work. We will never know what more might have happened with the fox, because my friend continued to focus on the songbirds and not the fox. That was her job, after all.

I think this is how wolves domesticated humans. Perhaps wild wolves approached humans because they had observed them for such a long time that they realized some individuals were not dangerous but rather nice and perhaps even fun. There is no better toy for a dog than a human who wants to play with them. Alternatively, humans could have been the last resort for wild wolves that were in bad shape—like the stork Hansi, who had followed the wrong friends into an area where he could not get any food. For Hansi, the only way to survive seemed to be by trying to approach the least dangerous human he could find: Grandma.

For a better future of our planet, I think we should all live up to these two situations. Either we help animals, or we play with them, or both. And in both situations, a relationship will develop and more and more animals will domesticate us.

18 Putin Invades Ukraine

AT THE END OF FEBRUARY 2022, we were anxiously waiting for the missing signature we needed so we could reprogram our tags after the Russian movie debacle. As soon as the signature arrived, we could get our global animal observation system back online. Then the world became a different place from one day to the next. At least for Europeans. Two generations had grown up with war as nothing more than stories about what our parents experienced as young kids or what our grandparents told us about the Great War. Wars like these were something we could not imagine ever happening again, but there we were in 2022, butting heads with another nation that was aggressively invading another country and killing and torturing people.

We were shocked and paralyzed. What about our friends in Russia? We knew they did not support this war, but we expected that they would be persecuted, punished, or maybe even deported or drafted if they spoke up. The time had come for democracies to once again take a stand against dictatorships in Europe. We were at a turning point in history, at least European history, and it had dire consequences for our observation system.

We followed the lead of the German space agency that carefully but clearly stated that bilateral projects in space between Germany and Russia were not to be continued. This handed the Russian head of Roscosmos, Dmitry Rogozin, the opportunity to make some truly bizarre statements, which he accompanied with some ludicrous animations that showed how he wanted to decouple all non-Russian modules from the ISS. All these modules rely heavily on Russian engines. Decoupling them would eventually put them into a downward spiral that would destroy the international modules of NASA, the European Space Agency, the Japan Aerospace Exploration Agency, and others, while the Russian part of the ISS could continue to fly on its own as a new Russian space station. ICARUS on the ISS was history.

Now we were happy we hadn't taught the Russians how to change the parameters for the ICARUS orbit propagation system because if they knew how to do that, they could continue to run the system. But they didn't have that information. The Russians also didn't know how to build or program the tags. The most they could do was try to get a bit more data from animals that had already been tagged because the ICARUS system was so resilient that it was still working even though we were no longer fine-tuning it. Over the coming few weeks or months the ICARUS receiver system would degrade until it was basically useless. ICARUS on the ISS had lost its wings—for the better, we felt, because trying to run ICARUS on the ISS had been like trying to look through a telescope smeared with butter. You couldn't see anything clearly no matter how hard you tried. ICARUS on the ISS would join other installations in the space cemetery in Baikonur.

The desperate phone calls we were getting from the Russian engineers, which we did not answer, told us that, apparently, they had realized these problems. We heard through the grapevine that the top level of Roscosmos was trying to run the ICARUS system to show how advanced their space engineering and space communication was. We felt deeply sorry for our engineering friends in Russia. It was not their fault that a dictatorship had taken control. The entirety of Russia was to remain virtually blank on the global animal movement map. This was tragic because owing to its sheer size, Russia is perhaps the most important country for tracking animals.

At the beginning of the war, it became clear that some higher-ups in the Russian military were searching for excuses should they decide to use biological or chemical weapons against their former friends and neighbors in Ukraine. This was nothing new, either for Russia, which used chemical weapons in Syria, or for Western nations looking for excuses to invade other countries or destroy certain industrial complexes of other countries where things were happening that they didn't like. Remember the search for weapons of mass destruction in Iraq? But the sheer stupidity of the argument brought forward by the Russians nevertheless astonished us.

Their accusations revolved around avian influenza. Avian influenza is a dangerous virus that circulates in wild birds, often waterfowl. Because of the interaction of wild birds with farm animals, the virus can jump to domestic ducks, geese, and chickens, and—very rarely—to people. Such spillovers are especially prevalent in areas around the great eastern lakes of China, such as Poyang Lake, and from there avian influenza can spread farther with other wild birds across Eurasia. Avian influenza has the potential to become a global pandemic with serious consequences for humans. A spillover of avian influenza to humans sometimes occurs in confined settings such as large-scale farming operations that raise chickens or ducks. This is one of the main reasons why being able to predict migration patterns of wild waterfowl across Eurasia is a high priority for all nations, and why we targeted wildfowl migration in Eurasia as one of our very first ICARUS projects. Let's all study this, we thought, and prevent the next major pandemic.

One of our graduate students had worked on avian influenza at a biological field station in Ukraine. When the facility was overrun by Russian forces, they went through the station records and saw that birds were being caught and tagged, and that samples had been taken to test for the presence or absence of antibodies against avian influenza viruses. What the Russian forces made of this was that Germany and some of its Western allies were actively investigating biological warfare systems in Ukraine and were planning to send infected birds from Ukraine into the Russian motherland.

This story made it all the way to the UN Security Council. The Swedish high military command and the German secret services got involved, as did the global press. The person who eventually stopped it was a top Russian general, when he suggested the avian virus had been genetically modified to affect only Slavic people. Thankfully, at this point it became obvious to everybody that the story was pure fiction. Apparently, the general had watched—and believed—the latest James Bond movie, *No Time to Die*, in which genetically modified material was programmed to affect certain people and not others. Even though the Russian general inadvertently exposed the story for what it was—a scare-mongering fairy tale—we were still accused of sending at least two birds (a goose and a crane) outfitted with ICARUS tags into Russia with the express purpose of spreading avian influenza. These two birds were supposedly the biological invasion that the Ukrainians had launched to hurt their Russian neighbors. It was a ridiculous caricature of the truth.

This fiasco finally convinced us that we would never again work with Russian engineers. We now needed to concentrate on our own ICARUS capacities, on even newer technologies, and on space assets such as small satellites over which we had full control. This would also help us make sure that animals were not used as a cover for crimes of war. To paraphrase the ancient Greek philosopher Plato, "necessity is the mother of invention." The war had been forced upon us, and we had to think differently, which is what we are doing now.

Currently, we are in the midst of the engineering design of our own ICARUS CubeSat. It will be launched in late 2024, and we will finally have a system that we control. The ICARUS tags are ready, sitting in the sun on my office windowsill, tanking up on sunlight and waiting for the day they can finally become wearables. There will be worlds to discover starting in 2025.

19. Cosmic Ideas From Aristotle to Humboldt

HUMANS HAVE ALWAYS HAD a desire to understand their place in the universe and see their connections with nature around them. In the Western tradition, understanding non-living natural phenomena such as gravity or orbits, as well as the interactions between such phenomena, began with the ancient Greeks about two thousand years ago. By contrast, trying to understand animals and their interactions only started in earnest with Charles Darwin's and Alfred Russel Wallace's ideas about evolution 165 years ago. And it was only ten to twenty years ago, during the biologging revolution of the early 2000s, that we began to get a good quantity of solid data to advance our understanding of how animals live their lives. This contrasts with Indigenous peoples, who for many thousands of years have carefully observed the natural surroundings on which they have depended for their survival and handed down their knowledge from generation to generation in oral form.

The ancient Greeks grasped that everything physical interacts with everything else through distant forces and that these forces hold the universe together. Things fall down onto the ground because there are forces making them fall. Today it is completely culturally ingrained in us that, for example, gravity exists, and even the most restrictive regimes in the world cannot argue that it doesn't. We all have to obey the law of gravity. But why, in the Western tradition, did philosophers and then scientists pay so little attention to what was going on with animals?

Perhaps animals were simply too familiar. Animals supplied food, clothes, warmth, and heartwarming (and sometimes terrifying) stories. Perhaps we only recognized the importance of this living universe once we became almost completely separated from it during the industrial revolution. This is not to say that the animal world was completely forgotten by all Western philosophers. It is just to say that major philosophical breakthroughs left out our connection with the animal side of nature. Clearly many other cultures and religions, such as Buddhism, were more tuned in intuitively to those aspects of the cosmos, but they were not an integral part of the beginning stages of what is known in physics as the grand unified (or grand unification) theory of the universe, which I think started with the early Greek philosophers searching for a single cause of our very existence. In a nutshell, this theory suggests that the major known physical forces were all one at the beginning of time.

Roughly two thousand years after the ancient Greeks started to formulate their thoughts about the physical properties of the world, the famed German naturalist Alexander von Humboldt and other global travelers started to incorporate newly discovered physical, chemical, and mathematical phenomena into a much more detailed canon of global knowledge. It was still possible to know just about everything there was to be known about physics and math at the time, at least for geniuses such as Humboldt. His book *Cosmos: A Sketch of a Physical Description of the Universe* is perhaps the last attempt in the Western canon to unify all of physics into a single volume.

Humboldt also tried to apply concepts from physics to the natural world. He believed there was not only a global physical connection of particles, but there was also a global natural connection of living things—that is to say, animals, as well as plants, fungi, and other life-forms. He believed that all moving particles interacted, whether they were animate or inanimate. But in Humboldt's time, it wasn't possible to quantify natural interactions around the globe. In contrast, in physics it was possible to assign general laws and mathematical formulas to the interactions of particles to capture the essence of what was going on. Take Humboldt's favorite scientific instrument, the sextant. You just needed to know the time and observe the relative position of the stars to know your own position on the globe. This amazing feat relied on the physical interrelationship of orbital systems, which in turn were based upon physical laws governing the entire cosmos. Or take an altimeter, to know how far above sea level you were, or a hygrometer, to know how humid it was. All these instruments measured physical properties that could be examined, captured, and understood. But there was no equivalent in the living world. This is perhaps why Humboldt referred to his book as a "physical description of the universe." It was clear to Humboldt that all things—physical, chemical, and biological—were interacting, and that the interacting whole was more than its parts, but unlike physics, where all these interactions could be quantified, there was as yet no way to formally capture animal behavior in predictive laws. All Humboldt could do for interactions between animals was describe what he observed.

At least for Western cultures, this meant that in all subsequent philosophical developments, the focus shifted away from nature, perhaps in a way that will be seen as one of the biggest failures of modern society during its entire cultural evolution. It was almost as if the benefits and functions of nature were considered a given. These functions existed anyway, no matter how poorly you treated nature and animals, which meant you did not need to worry about taking care of nature and animals. How else could you take the liberty of killing millions of bison for sport and profit, eradicate passenger pigeons, or transform awe-inspiring forests in the lowland tropics into pastures for the near-industrial production of hamburgers? And the way the West has raped and pillaged Indigenous cultures, and is still doing this, to destroy their knowledge of nature and extract natural resources, will collectively go down as perhaps the darkest chapter in human history.

The West has viewed the world as a place of infinite resources just waiting to be extracted for our benefit, but we are starting to understand collectively the folly of this approach. The difficulty has always been that laws that govern the animate world are more fluid and flexible than the laws of physics. You can bend them for a while, but in the end, the outcome is the same. If you are a human culture on an isolated island and cut down all your trees and live beyond your natural capabilities, nature will eventually hit back and you will have to leave your homeland, or die out and be nothing more than a little blip in the history of the planet. This is now the path we are on globally.

Perhaps the most important breakthrough in modern human history is the acknowledgment (which Indigenous cultures acknowledged long ago) that the laws of the animate world and nature are equally as strong as the laws of the inanimate world and physics, and that we need to obey them. This is not to say that Indigenous cultures do not change—and in some cases significantly alter—their environment, but it is to say that successful Indigenous cultures found ways to live together with nature for millennia, especially in ways where animals were treated with care and respect.

Bhutan, for instance, is a place where you are not allowed to mistreat animals. This cultural taboo links your family's history with your future. The worm that you smash with your ax or your shovel in the ground, or the fly you kill, could be your grandfather, or it could be you in the future. So, you better not do it. While this is obviously far-fetched in our current understanding of biology, the taboo provides a regulatory guideline, a cultural tradition, which is the simplest and most pervasive way to keep nature safe and people happy. And happiness is perhaps the most important good that can be achieved by anybody in our cosmos. Bhutan has kept its beauty, its nature, its potential for future development of the interactions between nature, animals, and humans, which makes it richer than most places these days. That is not to say that Bhutan is a paradise, but it is a shining example of where things have not gone as wrong as they have in Western cultures.

This might all sound like a doomsday story to you, but it is not. There is a way out. And it's a clear and straightforward way. It is the next chapter in human evolution. After the Anthropocene, when humans exploit and destroy nature, will come the Interspecies Age, when humans once again see themselves as part of nature. To describe the age that follows the Anthropocene as the Interspecies Age means that we will be considering other living species when we make decisions about what happens next on our planet, and that we are going to link the knowledge these other species have to our own knowledge. We are going to become partners. We are currently living through a most informative lesson in what we failed to understand about the cosmos. Growing awareness about the failures of industrialization is now providing us with excellent insights about how best to recalibrate our relationship with nature—and especially with animals.

Just fifty years ago, a silent revolution in telemetry started to give animals a voice on the planet. If we allow wild animals to become part of our daily lives—literally any animal—then we finally learn what Indigenous cultures and Bhutanese Buddhists already practice, and that is to link our own fate to that of animals, large and small, around the world. If any readers are not yet convinced that this is what we must do, let me highlight a few reasons why the Internet of Animals is important and why we should allow animals to be our teachers.

Let's first look at global pandemics. It is no longer necessary to remind people how traumatic a global pandemic can be. We were lucky that the coronavirus pandemic was not worse, that we quickly developed vaccines, and that, at least at the time of writing this book, it has not been a complete and utter disaster. Other animal-borne diseases such as SARS, MERS, and Ebola will come back, and will perhaps mutate and become much more dangerous in the future—or, one could say, in "their next round." If we had animal sentinels out there reporting back to us about diseases developing in animals brought into close proximity with people or farm animals, we might prevent future pandemics. No doubt there will be potential pandemics in the future because diseases are among the forces of nature, but what we can now do is make sure that we do everything to prevent the spillover from animals to humans. If we have a global system with thousands to tens of thousands of individual animals letting us know when diseases are spreading, then we can make sure that these diseases have only a minute chance of spilling over into our own lives.

And then we have anecdotes, children's songs, and folktales from around the world telling us that animals have a sixth sense that warns us when something big is happening in the environment. Animals have certainly not evolved to detect earthquakes or volcanic eruptions or other relatively rare natural catastrophes, but what they do have is a superb feeling for potentially dangerous situations. Fear is perhaps one of the most powerful driving forces in nature. Animals are constantly in danger of being eaten (except perhaps for a few that are top predators), of not finding enough to eat, of freezing to death, or of overheating. Thus, if collectively animals tell us that something has changed in the environment, that their world now feels even more dangerous, then we should listen. Picking up on danger is what they do best. Obviously we would want to corroborate with measurements of our own—satellite data, local sensor data, and the best of artificial intelligence models—but the natural intelligence of animals, the collective interaction of the most intelligent sensors we have on this planet, is perhaps our most important early warning system to help us anticipate natural catastrophes, at least on a local level.

The good thing is that with ICARUS, we can measure this kind of animal behavior anywhere on the planet in near real time. We can check in with animals to see how they are feeling at any given moment. And if they collectively agree that something is wrong, or dangerous, or threatening, this is what people intuitively accept as their sixth sense. It is what happens when a series of intelligent sensors are linked together. The nice thing is that the behavior that follows when these sensors are linked can be captured and quantified by the laws of physics and chemistry, which means we can analyze it and use it to predict what will happen in the future.

Imagine a flock of starlings dancing in an elaborate murmuration in the winter sky over a city such as Rome, Paris, or London. The starlings flock, in much the same way that fish like herring school, to avoid predators; in the starlings' case, these predators are mostly peregrine falcons. But while doing their aerial acrobatic dance, the starlings also sort out the heavy individuals from the light individuals. Think of a jar of sand when it is shaken. The heavier grains naturally gravitate to the bottom. Something similar might be happening with the starlings in their murmuration in the sky. The heavier—that is, the more well-fed—birds naturally gather at the bottom and are the first to land in the city trees to rest at night. The light starlings, those who did not find such good feeding grounds during the day, see where the stuffed starlings are resting and follow them the next morning to their feeding grounds to also, finally, stuff themselves. While this is all theory now, it shows how animal behavior can be based on—and therefore predicted by—simple physical rules. ICARUS tags will finally allow us to test these ideas and confirm how physical and natural laws interact. This is a huge step forward from Humboldt's day, when all he could do was observe and describe but not quantify animal behavior.

Animals are Earth's best biological observers because their lives depend on anticipating what will happen next. We have fleets of satellites watching the planet, but we still use guard dogs or sniffer dogs to help protect places or find missing people. We need local information. Nothing can replace eyes, ears, and noses on the ground, or the assessment of an ongoing situation by an intelligent being. This is what animals can do for us, if we give them the chance to tell us what they see and how they interpret their knowledge locally.

These warning systems could be as simple as many birds congregating where locusts are emerging out of desert soil that looks completely bare. Hordes of locusts might be on the verge of negatively impacting the lives of people who are already living on the edge. This is not to say we should then automatically spray all the locusts with pesticides and kill them, but advance warning would give us a chance to prepare an appropriate response. It would be similar with hurricanes or tornadoes. We could get people to take shelter until the disaster passes. In the old days in the U.S. Midwest, twisters were an unpredictable force of nature. Nowadays, with even a few minutes of warning, most people can shelter immediately and be safe.

Animals could also detect life-threatening conditions for humans and other animals alike. Vultures dying on a poacher-poisoned rhino carcass could alert us to a serious situation and prevent the slow and painful death of hundreds of other animals. Birds that can no longer form eggshells, such as peregrine falcons when the pesticide DDT was in common use, could tell us that there are disastrous chemicals in an environment that are not only threatening birds but also humans and whole communities in affected areas.

There are also so many relatively simple things animals could tell us just by being on the spot and doing the things that they do. Pigeons could measure traces of gas in cities. Outside the city, homing pigeons could measure wind speed, temperature, humidity, and turbulence in ways that would profoundly improve local weather forecasts. The nesting behavior of certain birds around the Pacific could be a superb predictor of upcoming El Niño events. The birds' behavior might even indicate the strength of what is to come. For example, these birds eat small fish whose population size is affected by variations in algal blooms at the onset of an El Niño event. Seabirds in the Gulf of Mexico could predict the annual harvest of anchovy or other fish much better than any human-made method or model because they have an inside track on the fishes' breeding successes and failures.

We are already seeing some results that point to how tracking animals can have immediate and concrete results for us humans. Take the example of Berta, the earthquake cow.

Cows as living earthquake sensors with ICARUS tags

20 Berta, the Earthquake Cow

WHEN I THINK BACK, I'm pretty sure my road to ICARUS began in 1988 when I was a university student studying biology in Munich. My best friend Bernd and I had decided to take a year off during our studies. We had read about Darwin and Humboldt and many of the other naturalists who wrote that they had been transformed by their travels. We thought, *Well, if it works for them, let's try it!* Our aim was to see in one year the maximum number of natural habitats that our planet hosts. At the same time, we wanted to understand the people that interacted with these habitats. How could we do this best?

My grandfather had just died, a traditional farmer who was very close with his animals and plants. He had left me many memories, a great deal of inspiration, and what, to me, was a large amount of cash: 15,000 German marks, which back then was about US$8,000. In our group of friends, we were discussing what we would do if we could just travel for a year. My plan was to burn through those 15,000 German marks. Hans, another friend, who spent days and months hang gliding through the Alps like a bird, came up with a simple solution. He said, "I would drive from Alaska to Tierra del Fuego. It's the longest stretch you can drive north–south from continent to continent."

Bernd and I thought that was a great idea. We started to make plans but soon realized there is in fact no road from North to South America. There had been attempts to build one through the Darién, a swampy area in eastern Panama, but even in its heyday that road had been no more than a muddy track, and by 1988 it had vanished back into the muck. If you wanted to drive that entire stretch, you would have to ship your vehicle from somewhere in Panama down to Colombia, which was a pretty daunting task potentially exceeding the 15,000 marks I had.

We decided we would concentrate on South America, one continent that seemed to have it all: fjords, salt lakes, the highest mountains, and the largest rainforest. I'm sure we were subconsciously influenced by Humboldt and Darwin, both of whom had traveled through the equatorial regions of the New World in the early decades after the United States of America was founded. Humboldt had explored the Casiquiare canal, a river that connects the continent's two major river basins, the Amazon and the Orinoco. Reading more about Humboldt's voyages, we really wanted to go to Venezuela to see Cueva del Guácharo, the oilbird cave.

While preparing for the trip, we still had to study, of course, one of those minor details and something you have to do to be eligible for academic positions later on. One morning I rode my motorcycle to Munich to sit my plant identification exam for a second time (which I failed again, by the way). A couple of blocks from the department building, I saw exactly what we needed for our travels: a long-wheel-base Toyota Land Cruiser Sahara. With its extra-large tires, extra-long chassis, and a place to sleep in the back, it was a vehicle that could take us anywhere.

The problem was, it wasn't mine. But as I was convinced this was precisely what we needed, I stopped. I wrote a nice note on a piece of paper I tore from one of my notebooks, and I stuck it under one of the windshield wipers. The note said: "Hi, You own a wonderful vehicle! This is exactly what we have been looking for, for a long time. We are taking a year off to drive through South America. If there's any chance, I would love to buy this car from you. Please consider selling it to us because it is the ideal vehicle for a one-year voyage from Venezuela to Tierra del Fuego."

I knew this was desperate and I knew the owner—probably a Munich yuppie—would read my note, tear it up, throw the pieces on the ground, stomp on them, and drive off. But that same night I got a phone call from the guy who owned the Land Cruiser. He said that he constantly found notes on his windshield from people wanting to buy his vehicle, but this was such a nice personal one, and we were about to do the one thing he had always wanted to do himself but couldn't, so he would sell us his vehicle for a mere 6,000 German marks.

We bought the Land Cruiser the next day and, with the help of my engineer brother, rebuilt it into a vehicle that would carry the three of us, sleep the three of us, and have enough storage space for equipment and supplies. As two of us were biologists, our main equipment was books—books about the animals and plants of South America and descriptions of various voyages, including the one by Humboldt. A few months later, the three of us were sitting in our vehicle at dusk just across from the Cueva del Guácharo in northern Venezuela waiting for the oilbirds to fly out.

As preparation for this awe-inspiring event, I had read Humboldt's description of his visit to the Cumaná area. He noted something that interested me even more than his description of the oilbirds. What Humboldt described about his visit here was that, together with his companion Aimé Bonpland, he had experienced an earthquake. He took all kinds of physical measurements, but even more importantly he talked to a lot of the local people, who told him that the animals tell you beforehand when an earthquake is about to happen. The entire passage about the way animals could predict an earthquake was somewhat vague. Maybe Humboldt did not want to stick his neck out too much on the subject, but there it was in black and white, a reference to the potential for using animals to forecast local earthquakes, written by one of my science heroes.

Humboldt's description stuck in my mind for twenty years. I came back to the oilbird cave in 2008 to fit some birds with the world's first remotely downloadable GPS data loggers. They had been designed by our young colleague Franz Kümmeth (who happened to be the son of my fifth-grade biology teacher) and engineered by e-obs, a biotelemetry company Franz had recently founded. What we discovered using these data loggers was that Humboldt's oilbirds were perhaps the most important seed dispersers in the South American rainforest. This astonishing fact was something Humboldt had no clue about. He had observed birds leaving the cave at dusk and returning at dawn. He assumed the oilbirds brought fruit back to the cave every morning. When they ate the fruit, they would spit the seeds out onto the ground. The seeds would then germinate and soon die in the dark cave. What our trackers revealed was that the birds didn't return to the cave every morning. Many days, they hung out in the forest instead and spat out most of the seeds there. The birds, as it turned out, were perfect seed dispersers for all the rainforest trees that had big fruit-covered seeds.

Luckily, our new tracking tags measured not only GPS position but also acceleration. Body acceleration lets you interpret an animal's behavior. Our unique and novel data could now be used to test the other idea Humboldt had heard from the Indigenous people at the time: namely, that animals could sense changes in their immediate surroundings before earthquakes happened. But how exactly could we test this? Naively, I thought I'd just go to those places where we knew animals had survived earthquakes because they took evasive action. Or to places where animals had successfully avoided the consequences of an earthquake, like a tsunami.

There had been reports that after the major 2004 tsunami in Indonesia that devastated Banda Aceh and many other places in the region, over 160,000 people had died. But on the little island of Simeulue, which was much closer to the epicenter of the tsunami and should have been devastated, only seven people had lost their lives. Everybody else had run up into the hills after being alerted by the chicken that flew around erratically and by water buffaloes that ran away from the ocean into the highlands. People relied on the signals from their animals and were saved.

There were also reports from Banda Aceh that a bunch of elephants had broken their chains and run away into the highlands just as the water buffaloes on Simeulue had done. The mahouts could not stop them and ran after them so as not to lose their animals, which was how they ended up being saved, as well. This gave me what I thought was an inspired idea. We would simply search for these elephants and their mahouts, give them our magic tracking tags, and if there were any more earthquakes in the area, we should see the elephants react before the earthquake or tsunami hit. At the same time, we would also tag and track wild elephants in the highlands, far from the dangers at the coast, to see if there was any difference in their reaction to or anticipation of an earthquake or a tsunami. I was proud of myself for this idea and approach, which is never a good sign.

It was not easy to find those elephants, get research permits, and source the right equipment to tag the elephants, but we pulled it off. It was wonderful to see how people lived with their elephants and how they worked together. This is not to justify capturing wild elephants and forcing them to work for people. That is horrible. But once you have a captive elephant, like a dog or a cow, I knew from my farmer grandfather that humans can have good relationships with their working animals. We also employed these working elephants to help us find forest elephants so we could anesthetize them and outfit them with tracking tags, as well.

Eighteen months later, after we had gone through all the incredible effort of tagging these tsunami elephants as well as some of their wild forest relatives, there had been no further earthquakes in the area. There was one earthquake some 300 miles (500 km) away in Indonesia, but 300 miles is a huge distance, and there was no accompanying tsunami. Although we learned nothing from our tagged elephants about predicting earthquakes, we did learn a whole lot of other things about them, such as their seasonal movements up and down from the ocean, as well as their interactions with each other. We also learned a few things about the behavior of some wild problem elephants that we tagged (elephants who were interacting with local farmers whose crops the elephants thought were theirs . . .).

So far our study of animals as earthquake forecasters had been a complete failure. Then I thought the next step might be to go into areas where a major earthquake had just happened and tag animals there to see whether and how they might anticipate the often massive aftershocks that occur after major earthquakes. The good thing about this idea was that aftershocks are somewhat predictable. We know they will happen and some of them can be quite strong. Not as strong as the major earthquake, but still, in a series of earthquakes where the main shock is magnitude 6 or 7, you can get a number of gradually diminishing aftershocks with magnitudes above 4 and as high as 5.

But to study this we had to move fast. We basically had to be in the area of the earthquake within two days of the main shock. And we also needed to have tagged animals within this two-day window. Shortly after the Banda Aceh expedition, I had prepared thirty GPS-acceleration logger tags programmed to gather exactly the information we needed: high-definition location information (GPS points) on the animals, as well as continuous acceleration data so we could understand their behavior every moment for at least two weeks. This was all the data we could be sure of recording without overloading the state-of-the-art biologgers at the time. And as this was all happening prior to the launch of ICARUS, we would have to manually download any data we gathered, but still, it was a start.

But this preparation still left me waiting for an earthquake and needing to be ready to fly there or drive there as soon as it happened. Every two weeks I had to recharge the batteries of the thirty data loggers and check that the clocks on them were synchronized and running. I also had to have on hand and maintain a wide range of wildlife wearables so I was ready to tag just about any kind of animals one could imagine. I had no idea where in the world an earthquake would happen next or what animals I would be able to find there and tag.

After about ten months of waiting, there was a major earthquake in Pakistan. I looked into this in detail and contacted some of the governmental relief agencies. But I realized very quickly that you can't go into an area where a major disaster has just happened and many people have died, and tell the survivors, "Oh, I'm sorry. I can't help you. But I would like you to help me tag your animals to see if they anticipate the aftershocks." This would not be the way to talk to people who had just lost their relatives, children, or parents and all their belongings and livelihoods. I wanted to help people in the future, but I could not do this by intruding on the grief of people in the present. I thought an earthquake in a European country might be a better bet, both because it would be easier to get to and because the overall situation would be easier to assess—possibly a country like Greece or Türkiye where earthquakes with strong aftershocks also happen.

In the summer of 2016, I went on holiday with my daughter to Scotland. Two days later there was the massive earthquake in Amatrice in the mountains of central Italy. I was briefly considering canceling our joint holiday, flying home directly, picking up the tags, and going to Amatrice, but I couldn't tell my daughter that an earthquake in Italy was more important to me than spending the holidays with her. And so, another opportunity passed me by. I was still charging the batteries on the tags every two weeks and checking that they were all in working order while waiting for the next earthquake. This was almost like the old days in Illinois waiting for the songbirds to take off at night, except by this time, I had been waiting for an earthquake for almost two long years after the expedition to Banda Aceh.

One night in the fall of 2016, two months after the Amatrice earthquake, I had just finished a nice dinner at an Italian restaurant in Konstanz with Uschi, ICARUS's project manager and now my partner, when we heard that there had been another earthquake in Italy, this time near the town of Visso, about an hour's drive north of Amatrice. We thought, *Let's drive there now!* But I had already had two glasses of wine, so there was to be no driving for me that night. It turned out that was for the best. The next day, the Italian president visited the area and everything was closed to general traffic anyway. Thankfully very few people died this time because after the Amatrice earthquake, a lot of the weaker buildings in town had either collapsed or been abandoned. Still, there was a massive amount of damage to livelihoods and houses and infrastructure.

We decided to start out on the ten-hour drive to Visso the next afternoon. Our plan was to arrive at 2:30 in the morning. We expected a lot of roads to be closed and thought that at that time of night, nobody would care if a car with German plates drove into the earthquake area. We were right. We took back roads and had to drive around some police roadblocks, but all the police officers were asleep at that time, probably because their day had been so exhausting. We parked in the main disaster area and slept in our car till dawn. Where to next?

First and foremost, we needed to scout out the exact location where we wanted to look for animals we could tag. The Amatrice epicenter, two months earlier, had been farther south. The central area of the current earthquake had apparently moved slightly north along the fault line, about 25 miles (40 km) away from Amatrice, toward Visso. We guesstimated that the aftershocks would happen to the north of the current epicenter. Apart from that, it was impossible to work right at the epicenter owing to the destruction of roads and buildings and basically everything else. We thought the best place for us to find animals, and perhaps even more importantly, people that would be willing to talk to us at such a moment of life-threatening stress, might be agritourism accommodations. Italy has a very nice system of former farms that have been transformed into vacation rentals where people can spend a cozy holiday in a luxuriously reappointed farmhouse in the countryside. Often these places still have some traditional animals, such as donkeys, cows, and horses.

But unfortunately, it was the end of October and most of those accommodations were closed for the season. We did find one where a custodian was still around but he chased us away, shouting, "Don't you know, we just had an earthquake here! What do you want?" We replied honestly, "Well, we would like to see whether the animals can anticipate the aftershocks." All the custodian said was "Leave right now. We don't need people like you here!" His reaction was totally understandable, but disappointing, as we had been working toward this moment for half a decade.

As we drove back to the main road, we passed through a tiny village with a lot of destroyed houses. One house looked like a farm shop. The windows were all smashed. The walls had major cracks. The tables with all the cheeses had fallen over and wheels of cheese were lying among broken glass. A young woman from the farm was apparently trying to clean up the mess—or at least inspecting the mess—so we stopped. Uschi, who's fluent in Italian, walked up and asked, "Can we buy some cheese?" The woman looked at her and probably was still in such a state of shock that she said, "Yes, we still have some not covered with all the broken glass. I can sell you some cheese."

This odd—and for the current situation, completely inadequate—conversation broke the ice. Uschi and the woman started talking. Uschi told her how we hoped the local animals might be able to forecast the aftershocks. The woman went to the farmhouse and asked her mother-in-law, who owned the farm, to come over. The old woman limped over and told us she had barely survived the main shock, as she had been thrown down a steep staircase in the old house when the earthquake hit. The shock waves from the earthquake were so strong that she couldn't hold on to the stair rail, and during the fall she had severely damaged her leg. Nevertheless, she was really interested in what Uschi had to say because her father knew—and had told her and she had told her children—that you have to watch the animals because they know when an earthquake is about to happen.

If the cows in the barn are covered in sweat and pacing nervously, then watch out.

On this occasion the family had had no warning. The earthquake had hit in the early morning hours and the cows had been fine when the family interacted with them at milking time the previous afternoon, perhaps twelve hours before the earthquake. The farmer told us she had already fed the cows a mixture of wine and sugar to calm them down. Such a treatment is important, she impressed upon us, because a cow stressed by an earthquake will show long-term after-effects such as giving very little or bad milk, having a stillbirth, or not conceiving a new calf for a while. This knowledge was also what her father had passed down to her. Her father was the one who knew from his grandfather that the quickest and best remedy for earthquake-stressed cows is to prepare red wine with sugar and let them drink it. They love it and it really calms them down. (I imagined it would do the same for me. I know this from the mulled wine that is offered at most holiday markets in Germany.)

Then the farmer led us to the barn and showed us her cows. The first one she pointed out was Berta, which she told us was the most sensitive cow on her farm. Berta was calming down after the wine remedy but still dripping with sweat, and many of the other cows were also covered in rivulets of sweat. While we were standing watching the cows, other family members came by and Uschi told them that we now had ways of monitoring animals continuously using the technology they all knew from their cell phones. The family immediately understood the value of such a monitoring system, especially for them locally, and allowed us, and even encouraged us, to tag their animals.

Finally, after six years of waiting and recharging tags every two weeks, we could attach the tags to animals. We didn't tag all the cows, because the farmer told us that not all of them were sensitive. But Berta, the lead cow, certainly got a data logger. We also tagged a few of the sheep, two of the four farm dogs, the five most sensitive chickens, the two most sensitive turkeys, and one of the four rabbits—all chosen by the family who told us that those were the individuals they would look at to see what might be happening. Twenty-eight years earlier, in Cumaná, Venezuela, I had sat outside the Cueva del Guácharo reading about Indigenous people who had looked to their animals for warnings about impending earthquakes. Now, in 2016, after many failed attempts, we were ready to observe this behavior for ourselves—or so we hoped.

We purchased all the cheese not sprinkled with broken glass, as well as some of the delicious local truffles that were still in their unbroken jars. Then we left the wonderful Angeli family to drive to the neighboring town of Norcia, which had suffered a lot of damage as it was closer to the epicenter of the main shock. We arranged to stay in the general area so we, too, could experience the aftershocks. We assumed it would be scary to feel the earth shake and wanted to have this experience because it would give us a sense of how it must feel to the animals. And would we, like some animals, get any sense that the aftershocks were about to happen?

We parked just outside the main town square in Norcia. Amazingly, one of the pizzerias situated just inside the ancient town wall had already reopened. The owners told us we were welcome to take a seat inside, but at the first sign of an earthquake we would have to run out and gather on the main square. We would be safe there, they assured us. The church had experienced many earthquakes and was still standing strong. We had a delicious, aftershock-free pasta dinner. Then we drove outside town to a place where sheep were grazing in a pasture. We thought we might hear the sheep bleating during the night if they felt an aftershock coming.

We didn't want to sleep inside a building, which did not feel at all safe right then, but we felt safe enough inside the institute's Volkswagen van and we had a peaceful night. There was no movement of the sheep, no sound from the sheep— and no aftershock. This struck us as odd, because major aftershocks usually follow within two days of an earthquake. We waited around all day. Nothing. Finally, on the evening of the second day, we decided that we had to go back to the institute to work. We drove to Germany, only to wake up the next morning to news not of an aftershock but of the start of a whole new series of earthquakes, the first of which had completely destroyed the church tower in Norcia, sending rubble tumbling down into the square where the restaurant owners had told us to gather just a couple of nights earlier.

Three weeks later we went back to the farm to download our data. We were unbelievably keen to see whether the animals on the Angeli farm really had anticipated this new earthquake. Not only did the Angeli family tell us that the animals had alerted them, but our data confirmed this. Hours before the main shock, chaos had erupted on the farm. There's usually some movement in the cow barn, but hours before the earthquake hit, the cows had frozen, which made the farm dogs extremely nervous. When there was dead silence in the barn, the dogs had started to bark wildly and run around, which in turn unnerved the sheep. Then the cows started to mill around in the barn, getting more and more worked up. For almost an hour, the activity of all the animals combined was about 50 percent higher than usual. If you know farm animals, then you know animals are maybe jointly active for a little bit when you feed them, but for them to be continuously active for almost an hour would be extremely unusual. All this activity preceded the major new earthquake by a number of hours.

Obviously, this was only one farm and only one series of earthquakes, and only seven out of the eight aftershocks with a magnitude larger than 4 were predicted by the animals. And yet, they had predicted these aftershocks hours before they happened. For the farmers and for us, this was a strong indication that the information gathered from Indigenous people and recorded by Humboldt some two hundred years ago could indeed be right: a group of animals can collectively indicate that an earthquake might be imminent. We later realized when taking a more detailed look at our data that the animals only anticipated earthquakes within about a 12-mile (20 km) radius. Intuitively, that made sense. Beyond that radius, an earthquake won't have much influence on animals.

We recharged the tags and left them on the same animals for almost another year. We later published the data in a well-respected scientific journal. As expected, we were heavily criticized for publishing such early findings. We did have a detailed timeline for animal activity before, during, and after a local earthquake. But we still had observed only one small farm in one earthquake area in one earthquake series. And, as there was no seismic equipment on the farm, we had to rely on data from the seismic station about 3 miles (5 km) away. Our study certainly wasn't perfect, but it was a good start. We would have been so much better off in our knowledge about animal predictions of earthquakes if we had had ICARUS and not just regular GPS tags on the animals back then, both on that farm in Italy and in other potential disaster areas around the world. We could potentially have watched the animal activities and behaviors ahead of many major natural disasters in near real time, and learned which animals could tell us what about impending disasters.

Two hundred years after Humboldt and twenty-eight years after I sat outside the cave in Venezuela, we had a sample size of precisely one. However, I am convinced that the earthquake in Italy has given us enough preliminary evidence to conduct more observations on animals to give those farmers who already know that they can rely upon their animals better tools to decipher what their animals are telling them at any time of the day or night. It's possible, of course, that animals might also act strangely for reasons other than an earthquake. Only time and a lot more research will give us the answers to that question.

21 The Internet of Animals

THE INTERNET OF ANIMALS is still in its infancy, but we have seen from the regular internet how powerful it can be to link information from around the globe. It is hard to imagine returning to the days when you couldn't easily communicate with your friends from anywhere on the planet, or when you had to physically search for references in libraries, or when you needed to travel the world to track down bits and pieces of information. Back in those days we ended up with stacks of paper on our desks almost like the papyrus scrolls in ancient Egypt—only more of them.

All these applications are extremely practical, but there is more. With the advent of algorithms that recognize behavioral patterns, we can now use the internet to predict human behavior. You could think of it as artificial intelligence recreating something we are already familiar with: our gut feelings. We don't know for sure what goes into our gut feelings, but what comes out are behavioral patterns that help inform us as to what we are likely to do next.

As our Internet of Animals gathers more data, it will create patterns in much the same way as the algorithms we are now using to predict human behavior. There are already enormous collections of animal data online, including data from natural history museums, data from spatial distributions of animals, and data on DNA sequences, diseases, or the conservation status of and threats to species and individuals. The key to the Internet of Animals is that we can now link this information to the real-time behavior of individual animals around the world. You end up with a system almost like today's traffic apps that link information-gathering devices (cell phones) to create a traffic pattern that is precise and available in real time. The amount, precision, and immediacy of the data combine to create an application that was unthinkable as recently as fifteen years ago. And from these data, you can predict traffic patterns in the future.

The power of the Internet of Animals is derived from the fact that the gathering, processing, and analysis of information will be distributed in a bottom-up way around the globe. People who study animals and can interpret their behavior will have the chance to input their information into a global tool accessible to all.

The true power of this developing Internet of Animals is currently still unimaginable, just as pedestrian navigational apps on cell phones had been to some participants at that meeting of the Bavarian Academy of Sciences not so many years ago. We could, for example, connect the Internet of Animals to other, human-made sensors, such as the radar sensors in modern cars, which could monitor wildlife along roads or even report on the presence (or absence) of insects. Home security and wildlife cameras are already used to record wildlife traveling through natural and unnatural habitats. By combining data from these sources with data from roaming pets and tagged wild animals, the Internet of Animals could alert residents and decision-makers to local wildlife issues. We also already use information from wildlife cameras to inform decisions about conserving species, but if we could combine these data with information from animal tags about poaching in real time, we could go so much further in our conservation efforts. There are other potential applications, as well. Tags on animals could supplement the sensors we already have for predicting natural disasters and extreme weather events, and disease monitoring via animal tags could help us stay one step ahead of what might otherwise turn into the next global pandemic.

Here's another way to visualize the Internet of Animals. Imagine planet Earth as a human body with each cell within it connected into an Internet of Cells. Immune cells detect pathogens and communicate their findings to the autonomic nervous system, which assesses the situation and decides on an appropriate response. This part of the nervous system does not necessarily know exactly what the problem is, but it knows something bad is happening and it initiates a response to neutralize the threat. This is how we can initially rely upon the Internet of Animals. We don't yet know all the details of the warning signs of individual animals, but like the immune cells in a body, the animals have a vital interest in keeping their host in good shape or they will all suffer. This is how we can work with nature. We have to keep her happy if we are to have a good life ourselves. We should honor and nurture our sentinels whether they are the immune cells within our body or the animals roaming our globe.

How do I think the Internet of Animals might develop? This is my vision. We would have what I will call a "lifecast" about our planet every day. It would be similar to a weather forecast, but on a global scale. The daily round-the-clock reports would livestream what is going on with life around our planet. A report might go something like this.

White storks and black kites have discovered another major out-break of desert locusts in southwestern Chad. Griffons in the central Himalayas are warning of an incoming storm, so Mount Everest climbing expeditions should remain at base camp. Good news from the seabirds. Gannets, frigatebirds, and sooty terns in Polynesia and around the Pacific Rim are indicating that this year will not be an El Niño year, as resources in the ocean appear to be excellent for breeding. Parrots, goats, foxes, bees, and snakes around Mount Pinatubo in the Philippines are engaging in their usual activities and squabbles, so nothing major is expected there in terms of volcanic activity today or tomorrow. The sea-eagles and porcupines around the Koryaksky volcano in Kamchatka are unusually active, though. Therefore, we expect a major volcanic outbreak there. All fifteen trappers and the two rangers in the area have been notified.

We interrupt the current report for an urgent message: Many animals on the small Indonesian island of Simeulue, in the east-ern Indian Ocean about 60 miles (100 km) from the west coast of Aceh province, indicate that a potentially major tsunami is on its way. We are sending out urgent warnings to people in the coastal areas of Simeulue and neighboring Aceh to follow the animals to higher ground.

Continuing the incoming reports. The domestic ducks at Poyang Lake in China are showing signs of disease and we are noticing migrating ducks coming through the area from around Hong Kong and from Indonesia and Taiwan. We have sent a team of medical technicians out to the lake to take cloacal swabs from the ducks to determine which strain of avian flu we are dealing with. We will give you a further update tomorrow and provide a prediction of where the migrating birds might spread the disease.

Further reports in the disease realm. Two of the straw-colored fruit bats migrating from Zambia to Ghana have remained in the eastern forests of the Democratic Republic of the Congo for the past few days. Their sensors indicate they currently have extremely high levels of antibody production, most likely to combat a variant of the Ebola virus. In conjunction with the data we have from many other fruit bats, some hornbills, and a few bonobos, we are now homing in on the area where the apparent host for this outbreak of Ebola resides. This appears to be a swampy and almost inaccessible section of the Congolese rainforest. We will install a local sentinel system there, employing tarantulas, lion tamarins, and cuckoos.

But there is more. The idea of an animal parliament comes from a children's book where representatives of all major animal species assemble after having decided that they, too, want to be heard. The good news is that a physical get-together is no longer necessary. Modern technology allows us to have a representative online parliamentary gathering of animals in which thousands of species and hundreds of thousands of individuals from all around the world can participate. What might discussions around the

day-to-day functions of an animal parliament look like?
Here is a possibility.

> *The United Nations has agreed to the request from the Animal*
> *Parliament today to establish a global banking system for animals.*
> *Individual animals can now hold bank accounts with balances*
> *based on the value of their services to their local community. Individ-*
> *ual animals are provided with an initial deposit into their accounts.*
> *Local people will get payouts from these accounts for protecting*
> *and generally safeguarding individual animals. The council of the*
> *Animal Parliament informs us that payments for the direct protec-*
> *tion of animals is finally possible because the health and well-being*
> *of these animals can be verified at any moment remotely anywhere*
> *on the globe.*

The idea of the bank accounts for individual animals
comes from the visionary thinker Jonathan Ledgard. As a
correspondent and editor for *The Economist* in Africa, he has
witnessed the problems of on-the-ground conservation for
decades. The principle behind his idea is simple. Bankers and
reinsurers strategize and make calculations based on much
longer-range thinking than politicians. These sectors have
their own armies of analysts and think tanks, and plan decades
or tens of decades ahead. They ask questions like "Will we
still have enough funding in the reinsurance system in fifty
or a hundred years to pay out retirement funds?" For this to
happen, we need nature to remain intact. If nature collapses,
everything collapses. The easiest, cheapest, and most dependable way to protect nature on site, anytime and anywhere in
the world, is to have animals roaming freely, safely, and happily.

To achieve this, one might be tempted to go the traditional route and create more national parks, guarding these

protected areas—sometimes by almost military means—against bad people and bad influences from the outside. National parks are a good thing in principle, and we certainly should keep them, but we also need a much more flexible, and thus better, way of protecting animals and the natural resources they rely on in areas of the world that are heavily impacted by human activities in our current era of the Anthropocene.

Many people around the world, especially in places where large animals still roam, such as in many parts of Africa, are without regular employment. The easiest way for them to survive is by "harvesting" the animals that live around them. But one could turn this situation around. These people could become the agents for a new, bottom-up global animal protection force. They could safeguard animal lives directly. Imagine a group of villagers living in a remote area of Zambia where giraffes still roam. These villagers might adopt a giraffe because for one thing, many of them like giraffes, and for another, they get paid. As long as "their" animals are doing okay, these local animal guardians get an immediate payout on their cell phones every day. And these days cell phones are a part of life just about everywhere in the world.

Up until now, the major obstacle for such a payout system has been verification. How can we verify for the banking and reinsurance sectors that an animal—wherever it is in the world—is safe and healthy? Well, ICARUS and the Internet of Animals will do the job. There would be constant communication from these animals. And similar to modern cell phones, you cannot steal, damage, or manipulate the data from the electronic tag on the animal. You cannot kill the animal you are supposed to be protecting and then pretend you, yourself, are a giraffe or a lion or a bat in order to continue getting payments from this animal's bank account. Therefore, if the animal you are supposed to protect is stressed or sick or moving on because it is unhappy, there is an immediate response in the reporting system of the Internet of Animals. Your daily payment stops and you lose your job and your livelihood. In contrast, if you and your local villagers ensure the wild animals—that, in a way, hired you to protect them—are happy, you will have a good and safe job for life.

The beautiful part of a global scheme like this is that the animals are now guardians of their own environment and can help create their own future. Animals need their natural surroundings to be healthy and safe. In this scheme, the animal protection payout system invests a small portion of global wealth in the direct happiness of animals guarded by local people. Once we allow animals to be in charge of their own fate, like in this animal banking scheme, it is amazing how many new and exciting options for conservation open up. All these options depend heavily on the Internet of Animals, that is to say, on the premise that individual animals can communicate directly with us.

Global banking is just one of the ideas about how an animal parliament could work to safeguard the interests of all animals. In a few years, or a decade or two, we will look back on the time before the animal parliament as we now look back on the time when women were not allowed to vote. The entire power system had to change (and in some places, still has to change) so we can become a true humankind. Interestingly, the last bastion of the old male-only voting scheme in Europe was the Appenzell, a canton in Switzerland, where women were finally given the right to vote in April 1990. Unimaginable, as we look back. It is not difficult to think back to an even darker time in the history of our planet, when mankind—and it was largely mankind then—decided some humans did not have the same rights as others. It was—and still is—a long, painful, slow process to establish that all humans are equal and that inalienable human rights are bestowed on every human on this planet. It is unfathomable that we are living in a time of globalization, and yet equal-rights issues are not resolved in every corner of the planet. But the trend toward improvement is unstoppable, and this trend touches all aspects of life. Overall, humankind is making progress toward taking care of other life-forms. It is no longer a question of if but of when this will come to pass.

This is not to say that animals will no longer face hardship or be hunted for food, but once they realize they are no longer being actively persecuted by humans or disadvantaged by decisions we make to further our own lives, their lives will be far less stressful. We know about this form of immediate stress reduction from human and animal stress studies. The most stressful situation is if you have no control. If you do have control and can live in your environment in the way nature intended, your stress levels will go way down, and you will be happier and healthier and live life to your full potential. Think about the arctic fox that started to play when it lived surrounded by abundance, or the rice rats in the Galápagos that explored my tent when it was clear to them that I would never threaten them.

In the Interspecies Age, we will for the first time be able to truly listen to animals and understand their needs and concerns in any corner of the planet by asking representatives of their kind carrying little electronic devices to tell us what's happening. Artificial intelligence will provide us with an even more advanced means of interpreting what animals are telling us, but it will not be absolutely necessary. We already have people who can interpret animal health and well-being and behavior: Indigenous people who have treasured and protected knowledge passed down over generations, despite the best efforts of Western colonizers to disrupt it; horse whisperers; dog whisperers; the Buddhists of Bhutan; and all the biologists and conservationists who care deeply about the animals they study. People have always had the ability to listen to the animals around them and interpret what they are saying, but now, for the first time in the history of the planet, we have an opportunity to do this on a planetary scale. Having this network of interconnected knowledge makes all the difference in the world. If you can sense the world on a global scale, you are bound to get a different quality of information and learn about aspects of life that you did not even know existed. And it is my goal to make global sensing an everyday and accepted part of our lives.

Epilogue: ICARUS Flies Faster, Farther

MANY PEOPLE HAVE TOLD ME that ICARUS is a bad name for a space system. But my answer is always the same: our Icarus is flying in low Earth orbit, far away from the sun. It is finally time for this iteration of Icarus to use its wings and truly fly.

When the Russian war on Ukraine started in spring 2022, those who had little faith in ICARUS declared once again that *nomen est omen*, the name says it all. However, war is also the mother of invention. In June 2023, our team began testing another experimental tracking system in space—signaling that our pioneering program for monitoring wildlife from space is set to continue.

The new ICARUS receiver will fly on a miniaturized satellite known as a CubeSat. For the first time, ICARUS will have complete coverage of the globe, including the polar zones. It will be able to detect birds, bats, marine reptiles, and land mammals anywhere on Earth. After an initial testing phase to lay the groundwork, the new, CubeSat-based operational system will commence data collection in fall 2024.

During the redesign process necessitated by the war, we have taken advantage of recent technological innovations. Measuring 4 inches (10 cm) on all sides and weighing about 4.5 pounds (2 kg), the new ICARUS system is smaller, more efficient, and more powerful than its predecessor on the International Space Station. Whereas the ISS infrastructure comprised a 10-foot (3 m) antenna and desktop-sized computer, the new ICARUS system has shrunk this to an 8-inch (20 cm) foldable antenna and a thumb-sized computer. The new ICARUS receiver will consume one-tenth of the energy while reading four times as many tags simultaneously, allowing researchers to download data and remotely reprogram tags faster. The system will be built by a Munich-based start-up that develops satellite tracking technology for research, agriculture, and logistics.

The 4-inch (10 cm) cubes that are assembled to create CubeSats are known as units (U). They can be combined to build CubeSats ranging from 1U to 16U in size. Being small and relatively inexpensive, CubeSats are increasingly being used by universities and institutes to launch research missions that would previously have been impossible. For its 2024 mission into space, ICARUS will be part of a larger 8U payload that will include several scientific experiments (the SeRANIS mission). Together, they will be hosted on a 16U CubeSat operated by another Munich-based start-up company.

The ICARUS host CubeSat, like the ISS and many other satellites, will be in low Earth orbit. At this comparatively short distance from Earth, the CubeSat can circumnavigate the planet fifteen times per day, meaning that it can cover any point on Earth's surface, and most of them at least every twenty-four hours. By contrast, the ISS does not cover polar regions north of southern Sweden or south of the southern tip of Chile. With its improved orbital path, the ICARUS receiver will be able to collect data from tags no matter where in the world an animal might be—in a desert, on a polar ice field, on the surface of the ocean, or up in the sky. With this increased coverage, we have strengthened our mission of global monitoring of animal biodiversity, and we will also be able to monitor the polar regions that are the most at risk as the climate changes.

With a single ICARUS receiver in space, data will be read once a day, providing researchers with regular updates on animal behavior. In the future, the ICARUS system will be extended to a network of receivers on satellites, which will increase the number of daily data readouts. Plans are currently underway for a second ICARUS CubeSat receiver to be launched in 2025, and a third in 2026. The goal is to have enough receivers in space to deliver near real-time data transmission—a result that will have important implications for conservation, as real-time data will give conservation managers around the globe the ability to safeguard biodiversity much more efficiently.

We are also working to fundamentally redesign the ICARUS tags themselves. Currently, our ICARUS tags—which weigh a fraction of an ounce—record an animal's GPS location, movements, and surrounding environment, such as temperature, humidity, and pressure. The new tags add two-way communication, so that tags can be reprogrammed by instructions sent remotely via space. Additionally, onboard artificial intelligence can help "decide" which data should be collected, based on the behavior of the animal. We will also shrink the size and weight of our tags to about half their current values.

We have spent the year since the war robbed us of our space receiver optimizing tag design and incorporating advances in AI. What we have now is akin to a quantum leap forward: a system that better helps the animals tell us their most important stories.

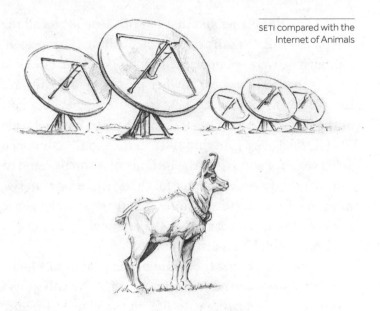

SETI compared with the
Internet of Animals

Afterword: A Glance Into a Bright Future

MANY PEOPLE THOUGHT Sputnik would usher in a new age of space exploration. And, indeed, it did. But as is the case with most revolutions, true change is not what happens most prominently in front of everybody's eyes. True change, and the true power of revolutions, is often what happens in some little corner where the forces of the revolution are harnessed in a much more careful and subtle way. This was the case with animal telemetry.

The industrial revolution, the space revolution, and the digital revolution now provide the tools for humankind to truly reflect upon life on the planet—to understand that we are sharing this planet and that our life depends on all the other life-forms. We can't rule the planet from the top down as though we were self-appointed dictators for life. The good thing is that all of us benefit when we listen to the diverse needs of life on the planet. These ideas are not new, but now for the first time we can really apply them on a global scale. The knowledge we gain through listening to the combined global voices of animals in the Internet of Animals could be a much more powerful impetus for change than anything we may learn from the search for extraterrestrial intelligence, aka SETI—an endeavor that has taken up so much of our time, energy, and financial resources.

We now realize that the entire planet is akin to a living organism and we can't be our own cancer, or we will vanish. Our new way of dealing with life on the planet also goes way beyond merely caring for pets or livestock. It includes paying attention to the voices and needs of all life-forms. In some wildly twisted way, we are now able to use the formerly destructive forces of industrialization—those that separated us from animal life—to bring us back to a global understanding of just how interconnected all life is on planet Earth.

It is May 4, 2055. My daughter and her children are watching the evening news and the weather forecast for tomorrow. Then comes the lifecast, which on this day is focusing on animal migration and recommending that people tune in to reports from three global ambassadors: the bar-tailed godwit Theunis, who is taking off on his eight-day nonstop flight from New Zealand to Alaska; the bonobo Nikita, who is staying put and teaching her children how to find the best fruit in the Congolese rainforest; and the death's-head hawk moth Celine, who has just started to fly from Lago Maggiore in Italy toward Konstanz in Germany, straight over the Alps. My grandchildren, like millions of other children around the world, tune their devices to catch up on the latest news from these global ambassadors and then check in on the animals they have signed up to protect remotely: a baby rhino and a riverine rabbit.

Significant and sustainable change in human cultures around the planet often comes quickly, and the new normal tends to be the mindset of the majority in the end. Some of us will remember how this worked for leaded gasoline or for smoking in planes and, later, restaurants. Many commercial airplanes still have ashtrays for cigarettes in the restrooms. And what about telephone booths? My kids asked me once what this yellow box standing on the sidewalk was good for. It took only a single generation to forget or erase from our memories that these things had once been the norm.

Throughout human history, such periods of rapid societal change often happened when major religious transformations occurred or when a single prominent figure on a global mission started to influence the population at large. George Swenson, my old radio astronomy friend, was convinced that if successful, the search for extraterrestrial intelligence would provide just such an impetus for major change in the human culture. Imagine what might happen if humans made contact with extraterrestrials. Connecting with other intelligent life-forms in the universe would certainly put our human existence into perspective. We would no longer be the chosen ones; instead, we would merely be one of many.

George always impressed upon me that the search for extraterrestrial intelligence was an important endeavor for humankind, and if we were ever successful, it would be one of the most transformative single events in our history. But would it really? Any messages we might receive from somewhere in outer space, albeit incredibly interesting, would be massively out of date. The problem is one that we haven't solved yet, and one that also prevents us from doing true human space exploration. In our current understanding of how physics and the cosmos work, time and space are interchangeable. A message that arrives via radio waves (that is to say, at the speed of light, which is approximately 186,000 miles per second or 300 million m/s) from somewhere far away will have traveled a huge distance and that would have taken a huge amount of time. A thousand light years would be no distance at all for such a message to travel. And our response to a message that is one thousand years old would then take another thousand years to reach the entity that sent the message in the first place. To say nothing of the fact that we would have to send our response to arrive at exactly the right spot in the universe.

Just these few thoughts should suffice to make us realize that communication with extraterrestrial intelligence out there in the universe is something that is unlikely to happen until we have new information about the physical principles of space-time travel. (And it would similarly be prudent for major human spaceflight programs to first invent faster transport systems before embarking on near-endless human space travel times to even the closest of our uninhabitable neighboring planets.)

Overall, I think we can do better than look to outer space for answers or groundbreaking new ideas. What we have learned by listening to animals is that they often have a totally different perspective on life on the planet than we humans do. Listening to animals might actually change our human way of thinking more profoundly than any unlikely message from outer space. As we start receiving messages from animals and truly listen to them, humans would also be more disposed to losing their culturally ingrained perception (at least in the Western world) that they are the God-given pinnacle of all life-forms. Those of us not already on board with the concept could finally accept—and this would need to be culturally ingrained—that we are just one of many life-forms existing in our universe (our universe being either the entirety of our own planet or the greater universe as a whole). We are quickly running out of options. The attitude that brought us to the Anthropocene is now endangering our global future—though not the future of life on the planet itself, as many doomsayers wrongly suggest. Earth has witnessed many catastrophes, and some forms of life have always survived and prospered over the eons of our planet's existence. It is our own future that we are putting on the line.

Listening to the transformative intelligence of animals also transcends the problem of space and time that is inherent in the search for extraterrestrial intelligence. When tuning in to the Internet of Animals, we are getting the messages from animals right now. And along with the current messages, we are also receiving the collective knowledge animals have accumulated over the course of their evolution, allowing us access to a vast store of ancient wisdom.

Throughout this book, I have tried to highlight the conceptual changes I can see happening when we listen to animals. Some are simple things, such as understanding that animals know their environment and their homes much better than we think. The little rice rat on Isla Santa Fé in the Galápagos returned to my campsite more quickly than I ever would have imagined, for example. But we have also learned that animals are apparently constantly communicating with each other. Berta the earthquake cow showed us that she is in constant communication with all the other animals on the farm. And we think the starlings learn from each other during their murmurations which birds have found the best feeding sites in winter. These three stories taken together give us an insight into the existence of the so-called sixth sense of animals, which is nothing but an example of how intelligent sensors acting together offer more information than individual sensors on their own. Little songbirds listening to all the other birds on the migration highway is another simple but powerful example of how such inter-animal communication systems transfer knowledge to individuals, knowledge they would not have had without the collective.

But maybe the most important lesson we can learn from the intelligence of animals is that they still consider most of us colonialists—intruders who want to take over their territory—while we, in contrast, believe that we are the ones moving things forward and making things happen. Until I learned about the stork Hansi or the playful arctic fox, both animals well on their way to domesticating humans, I was convinced that it was humans who domesticated animals. This is what the biology textbooks of learned old white men had imprinted upon me. Letting animals tell us their side of the story certainly puts that limited worldview into perspective.

Perhaps the key transformative principle we need to accept from the collective intelligence of animals, one already long accepted by most Indigenous cultures, is that life on Earth is entirely interconnected. Once we accept this idea and make it our own, we will be able to safely live on planet Earth and look forward to a bright future. Obviously, this one planet has always been our home, but now we need to embrace and act on this knowledge. In the past there were so few people and so much space that we didn't have to manage our global home as one unit. It was still okay for all the local villages, later kingdoms, and nowadays nations to live by their own rules and procedures, which often conflicted with the global needs of life. A focus on local systems was possible because local decisions had a minimal impact on the global system. But now, in the Anthropocene, our human influence on life as a whole has changed not only qualitatively but also quantitatively. Everything we humans do now affects the rest of the planet directly. We can no longer hide in our local valley or shire and pretend that everything is good here. If we continue to do that, we will follow the fate of earlier populations that destroyed their natural resources: we will vanish.

The good news is that signs of change are on the horizon. Despite all the daily and often horrible news, global ethical and legal codes for planetary life are already being formulated and work in this area is accelerating. Working toward global climate solutions is a good first practical test of what is to come. The actions toward global biodiversity conservation will bring nations together even more. We are already on a solid path toward a globalized knowledge of the physical properties of our planet. We already share data on weather and climate, as well as remote sensing data that include plant life as the basis for animal life. With a delay of about two to four decades, we are now also on a good path toward globalizing knowledge about animal life. And even better, we now have the technological means to learn directly from the global collective of animals.

It is within our reach to communicate with animals, and the messages we receive from them have the potential to change our human trajectory into the future for the better. We can use the technological advances of the industrial revolution that brought us to the brink of our own extinction to now bring us back toward the direct understanding of, and communication with, all the other life-forms on the planet. These new abilities will give us a thorough understanding of our own role and position within the system of life. We will no longer be the conquerors of other human cultures and the animal kingdom but rather equal partners. Partners that talk to each other and listen to each other.

Successful Indigenous cultures have a long history of living within the boundaries that nature sets for us. Our task today is to become a global Indigenous nation. I don't mean the kind of globalization that erases differences and makes everything the same, but the kind of globalization that allows each community of people and animals to live in a way that is most appropriate for and best suits their local environment. This includes maintaining and enhancing all the local and regional differences that we are proud of.

How lucky we are to live in this very moment. We have the chance to become the first global nation. All we need to do is adapt to the natural rules of our planet. This is our path from the Anthropocene into the Interspecies Age and the way humankind can thrive.

Acknowledgments

AS THIS BOOK REPRESENTS an amalgamation of experiences from various stages in my adult life, it is impossible for me to thank everybody who influenced me and my thinking. I name several colleagues and friends, but many more remain anonymous. Whether they appear as characters or not, I would like to thank all of them tremendously for their input, their caring, and their criticism. I should also acknowledge that sometimes bad teachers influence you more than good ones by demonstrating what not to do.

As a scientist, I believe that human thought processes, including the thinking that went into this book, are an integral part of our makeup and therefore, like any behavior, thought processes can be analyzed to shed light on behavior in the past and help predict behavior in the future. To do this for any individual—whether human or animal—you need to address four major questions formulated by the great animal behaviorist Niko Tinbergen: you need to know what the individual's family background was like (phylogeny), what its childhood and upbringing were like (ontogeny), the stimuli it is exposed to on a daily basis (causation), and its ability to survive, including near misses (adaptation). I'd like to apply these four questions to my life and acknowledge the contributions people made to help answer each one.

First, phylogeny. Where do my thought processes come from? I am indebted to my father's family, who were, in succession, immigrants, then refugees, re-immigrants, re-refugees, and finally immigrants again, all within two generations. On my mother's side, I am grateful for the collective knowledge handed down in a family that resided in a village with fewer than fifty inhabitants since its founding in Bavaria in 1242.

Second, ontogeny. What were my childhood and upbringing like? My father carefully demonstrated to me what you can learn from individual animals when you tune in to them, and how to handle a baby barn swallow or observe a badger family in the middle of the night. My grandfather and my mom told me how important it was to stick to your guns, even if five Nazi guns were pointed at your body. My teachers and friends during my Catholic school years taught me the principles of philosophy, but also that you should never give in to nonsense and accepted practices within your own institution.

I had to largely skip university as soon as I could because I wanted to travel all the continents and the in-betweens before getting to work. Traveling South America with my friend Bernhard Gall and my brother, Thomas, was certainly a life-changing experience. Working nights and weekends at my beloved South Tyrolian bar with Hubert and Carla Schwienbacher taught me about the importance of pampering your customers, and about the beauty and fun of constant innovation and the satisfaction of helping others. This job provided me with the means to learn how to fly like a bird with my friends Hans Eckl and Axel Rabeneck. I thank my brother, Thomas, and my friends Hans Eckl and Bernhard Gall for sharing my hunger for travel—especially when our level of hunger could be measured by the depth of the trash cans we were willing to dig down into for food. The biology field projects on marmots in the Alps, sea lions and marine iguanas in the Galápagos, and crabs in Jamaica, as well as the exposure to discussions in the Wickler Café with members of the Department of Behavioral Ecology at the Max Planck Institute in Seewiesen, were all instrumental in shaping how I think about the world.

Third, causation. What stimuli am I exposed to every day? An academic life, most of the time, is like being self-employed. It's free, it's beautiful, and you play your butt off! Before I moved back to the Old World, my friends and colleagues at the University of Washington, the Smithsonian Tropical Research Institute, the University of Illinois at Urbana-Champaign, and Princeton University taught me invaluable lessons. Jointly, we prepared the ICARUS white paper, the document that outlined the beginnings of the Internet of Animals.

Back in Germany, living close to gorgeous Lake Constance in the middle of Europe, culture became a more important part of my life. The deep insights of my friends and colleagues at the Max Planck Institute for Ornithology helped us found the Max Planck Institute of Animal Behavior, a fun and highly productive place to dive deeper into the world of animals. The spiritual teachings by my Bhutanese colleagues Sherub and Nawang Norbu were a particular highlight. I extend special thanks to friends in the greater Konstanz area—Silvia and Meinrad Arnold, Gabriela and Hannes von Witzleben, Claudia Saunders, and Jan Dodel and Eva Hoeffmann Dodel—for thoughtful insights beyond the facts of science. And particular thanks to my beloved partner, Uschi Müller, the ever-powerful Rhinelander, who keeps reminding me that even though the challenges are constant, everything works out, particularly when you are having fun.

Fourth, adaptation. How have I managed to survive and what near misses have I had along the way? I am indebted to my colleagues at the Charles Darwin Research Station for pumping 25 U.S. pints (12 L) of saline through my body in a single day during a cholera outbreak. Thanks to Bernhard and Thomas for walking me down Mount Roraima in Venezuela when I had malaria. Bernhard was also by my side when we had to drive through roadblocks near Tingo María, Peru, to avoid a Sendero Luminoso killing squad. Ela Hau helped me stanch the flow of blood from my femoral artery in the Galápagos. And thanks to the storks that avoided my Cessna plane during a low-level stall when a helicopter filming us flew right in front of me. These and a few other incidents taught me to enjoy life, live like a bird, and be thankful for every day.

As I mentioned at the outset, there are too many people who taught me essential lessons throughout my life for me to name them all, but I would like to thank each and every one of them with this book.

And, lastly, I would like to thank the team at Greystone for making this book happen. Thank you to Rob Sanders for encouraging me to write it, to Dawn Loewen for her meticulous copy editing, and to the design team for such a beautiful package. Special thanks to Jane Billinghurst for excellent editing and a most wonderful collaboration throughout the creative process.

Appendix:
Initial ICARUS Projects

THESE INITIAL PROJECTS are only the teaser, the bare beginning of a new age of exploration and knowledge.

The Life of Sooty Terns

*Sooty terns on Ascension Island and in Polynesia and the
Seychelles*

Sooty terns (*Onychoprion fuscatus*) are found in tropical and
subtropical oceans around the globe. The striking black-and-
white birds spend most of their lives at sea, where they feed
on small fish, squid, and crabs. They only come ashore to
breed, when they form huge colonies of up to a million individuals. While almost all populations breed at twelve-month
intervals, the birds on Ascension Island in the Atlantic do so
every nine months. What this difference is all about is still
unclear. With the ICARUS transmitters, researchers want
to solve this mystery and find out how sooty terns navigate
over the vast oceans and find their way back to their home
island. They also want to know if young terns really do live
exclusively in the air for four years and where they fly to.

Global Songbird Migration

Blackbirds and thrushes in Germany, Russia, North America, and Tibet

Billions of songbirds migrate between continents twice every year. However, not all migratory birds set off on a journey every year. Some simply stay where they are. Such partial migration represents a transition point between migration and sedentariness. Why some birds of a species migrate and others remain at their place of birth is a mystery, as suitable technology for following songbirds throughout the year has not previously been available. Knowledge about partial migration—where only part of the population migrates while the rest remain where they are—helps us to understand why birds migrate. It provides insights into how environmental conditions during the nonbreeding season lead to adjustment strategies, and the extent to which global environmental changes, such as climate change or changes to land use (for example, urbanization), generally influence migration strategies.

Songbirds provide essential ecosystem services for humans—such as insect control—but over the past twenty years, their numbers have decreased by 30 percent. How to protect them is not clear yet. The ICARUS team plans to track five thousand blackbirds and thrushes in Eurasia, Russia, and the Americas as a pilot project to start understanding where these songbirds live and die, and how to protect them. The goal is also to discover how flexible the birds' decision-making on migratory flights is when they live in polar, temperate, or Mediterranean regions. The researchers hope to understand how migratory birds overcome the challenges of their environment and whether they could respond quickly enough to changes such as climate change or urbanization.

Juvenile Travels

Bears on Kamchatka, pumas in Central America, turtles in Florida, giant land tortoises in the Galápagos, and cheetahs in Namibia

The most difficult time for most animals is when they move away from their birth area. In many species, the whereabouts of these juveniles and their travel routes are unknown. Most individuals die during this difficult time. ICARUS plans to study these lost years with long-lasting ear tags in mammals, or with small solar tags in sea turtles, giant land tortoises, and seabirds. The project will fill numerous gaps in knowledge about the lives of young animals and facilitate the protection of endangered species.

Great Ape Help Line

Orangutans in Southeast Asia

Although humans' closest animal relatives are highly endangered in the wild, in places where we study them continuously, they usually reproduce and live well. At the same time, thousands of orphaned or rehabilitated great apes are waiting to be released back into the wild; however, they first need to be properly protected using wearable tags. ICARUS plans to optimize wearable communication anklets for great apes to enable them to call for help. Ethical considerations are the centerpiece of this effort.

Human and Animal Movements

Livestock in Bhutan, the Sahel, East Africa, and the drylands of Asia

Throughout history, nomadic herders have moved around with their livestock. In some remote areas in the world, we can still witness these joint migrations. Who guides whom, who learns from whom? Scientists will study joint herder-livestock movements in remote areas.

Animal Rangers
Large mammals in South Africa's Kruger National Park and in Kenya

Rangers are important for the protection of wildlife in most areas of the world, guarding against poachers and other threats. Rangers cannot be out in the wild all the time, but fellow animals can. Animal rangers can help warn against predators as well as against human poachers. Researchers have developed ways to learn from the collective of animals to understand when poachers are on the move.

Pandemic Alerts
Bats in Zambia, Ghana, and Rwanda

Humans are encroaching on wildlife everywhere. This can result in pathogens crossing the species barrier and jumping from animals to humans. Bats are usually determined to be the culprits, but most often they are falsely accused. We need to find out more about their interactions with other wildlife and humans, and where these interactions occur, to shed light on this major link between human and eco-system health.

Protecting the Flight Paths of Wading Birds
Wading birds in Australia and East Asia

The East Asia–Australia flight path is one of the most threatened routes in the world for migratory birds. Gathering points in Asia are increasingly falling victim to urban development. Most migratory bird species are in decline, and many are now threatened. Wading birds are also being affected. This category includes well-known species such as curlews, plovers, and arctic terns. ICARUS will help shed light upon the migratory routes of wading birds and identify the main gathering points in order to protect them.

The Role of Fruit Bats in Ecosystems
Fruit bats in West Africa

Many fruit bats are extremely mobile. On their nightly journey from their sleeping to eating locations and back, some cover over 60 miles (100 km) per night. Because they carry pollen and seeds over large distances, they play a key role in the pollination and proliferation of plants, and thus in the natural renewal of forests and in human nutrition. They are also increasingly mentioned in the context of disease, although evidence for this is in most cases indirect via antibodies or DNA fragments. We contribute to a more holistic approach for the study of viruses in fruit bats by tracking their movement and ecology to find where contact with diseases occurs and to understand the potential for transmission or spread of diseases in those places by the true disease hosts, which are not the fruit bats. To understand the bats' role as keystone species in the ecosystem, detailed knowledge of their movement behavior is important. Researchers also know little about the impact of hunting and habitat destruction on their population numbers.

Straw-colored fruit bats (*Eidolon helvum*) are the most common fruit bats in Africa and form large seasonal aggregations. Yet little is known about the connectivity of these colonies, what predicts their presence in different locations across Africa, or what role they have in ecosystems—all questions that can be answered only by following individuals. We will attach ICARUS tags to straw-colored fruit bats to monitor their migration across Africa. The tags will regularly communicate with and upload data to ICARUS satellites, allowing high-resolution GPS tracking of animals in remote areas.

Young Predators in Unfamiliar Terrain
Jaguars in Central and South America

When young predators leave their mother, they also have to look for a new territory. Various populations of a species thus encounter one another and can even establish a completely new population in the process. Juveniles move around extensively, often passing through areas humans have settled in or developed. The young animals are inexperienced. They are familiar only with their home territory and have little knowledge of how to avoid dangers in such environments. Researchers hope to deploy the ICARUS transmitters to identify the routes of young jaguars in landscapes dissected by roads and urban settlement. This will also provide insight into the criteria jaguars use when choosing their travel routes. Proximity sensors installed in the transmitters will document how the roaming animals react to conspecifics, especially around human settlements that represent danger for them.

Traveling Ducks
Ducks in Siberia

Billions of ducks breed in Siberia and migrate southward to overwinter in the tropical regions of Africa, Asia, or Southeast Asia. Northern pintails (*Anas acuta*), for example, set off on their journeys at the beginning and end of the breeding season, whereas mallards (*Anas platyrhynchos*) appear to move back and forth between various locations with much greater flexibility. Little knowledge exists about which routes the species take, where they make transit stops, and how the various duck populations are connected with one another. Such information is nevertheless becoming increasingly important in order to protect the animals and their habitats, on one hand, but also to prevent the spread of infectious diseases, on the other. This is because ducks are potential carriers and vectors of a high number of infectious pathogens, such as the bird flu virus and antibiotic-resistant bacteria. ICARUS scientists intend to equip individual birds of the two most common duck species with transmitters at various locations from which birds fly north toward Siberia. The researchers will also test the birds for a series of pathogens. Furthermore, they will measure their body temperature to know when they develop a fever and their body acceleration to provide an indication of their energy expenditure.

Orphaned Bears
Brown bears in Romania, the U.S., and Canada

How do young bears raised by hand cope when they are released back into the wild? The ICARUS transmitters allow researchers to track the movements of young bears and evaluate the success of reintroduction programs.

Flying to Africa Alone
Cuckoos in Europe and Africa

The migration of songbirds is one of the most remarkable natural phenomena. It is incredible how birds undertake their journeys and arrive together at the same location over many years. Even young birds cover thousands of miles alone and find wintering grounds they have never visited before. Many details about bird migration remain unclear, as following bird movements over long distances is difficult. ICARUS enables scientists to track various species of cuckoo on flights where young birds undertake their journeys completely independently, without any experienced conspecifics. This gives researchers an insight into what the birds use for orientation and how their routes have emerged over the course of evolution. They can also evaluate whether young cuckoos can adapt their flight direction if they have strayed from the path and from what age they are able to do so. In future, scientists plan to equip small songbird species with even lighter transmitters to establish whether the results can also be transferred to other migratory birds. In view of the short succession of generations of small birds, it is easier to analyze the consequences of environmental changes on survival and reproductive capability.

Protecting Antelopes
Saiga antelopes in Kazakhstan and Mongolia

Saiga antelopes (*Saiga tatarica*) almost became extinct in the 1920s. The population then recovered to reach almost two million animals. However, Saiga antelope numbers have again fallen sharply in recent years owing to intensive hunting and habitat loss. The animals have also been hit by a deadly infectious disease since 2015. ICARUS aims to foster the protection of these critically endangered antelopes. The data will indicate which regions are of importance to their survival. This will allow designated protected areas to be created, where the animals can take refuge in the rutting season, for the birth of their offspring, and during migratory movement.

A Family Affair
The world's fifteen crane species

In contrast to songbirds, families of cranes stay together during their first migration and sometimes for longer. Scientists suspect that parent cranes pass on their knowledge about bird migration and feeding places to the young animals. The route and time of migration are constant among individual birds; however, in many species, it has not yet been established whether the young cranes select the same flight path and the same time for their journey as their parents after separating from them. ICARUS researchers want to observe the flight paths of crane families throughout their entire lifetimes. They are attaching transmitters to various species of cranes in their breeding grounds and comparing the movements of family members on their first migratory journeys. They are also exploring which factors influence the timing of migration and which animals lead their flights, including their incredible flights across the Himalayas. They also want to find out when and how young cranes separate from their parents and how the family members find one another again after separation. The results will help to provide endangered crane species with more effective protection.

Index